Analyzing

Complex Adaptive Systems

Basic Quantitative Approach

Shahin A. Shayan

Copyright © 2021 Shahin A. Shayan
Amazon Publishing

All rights reserved. No part of this publication may be reproduced, stored in a retrieval system, or transmitted, in any form or by any means, without prior permission in writing from the author.

Dedicated to my family and all those who have contributed to the accumulation of knowledge and understanding throughout history

Humans are members of a whole
In creation, of one essence and soul

If one member is afflicted with pain
Other members uneasy will remain

If you have no sympathy for human pain
The name of human, you cannot retain

From Saadi Shirazi[1]

Ut Omnes Unum Sint[2]

From Latin

Nonlinear interactions between agents cause
Complexity, and learning plus cognition causes
Adaptive behavior.

From This Book

---

[1] **Saadi Shirazi**. A medieval Persian poet who lived between the years 1210-1292 AD. A highly moral and socially conscious poet with two literary masterpieces called the Bustan and Golestan – Also see Wikipedia.

[2] In Latin means, we are all one.

Table of Contents

Preface	5
Introduction	7
Networks	11
Systems	45
Complex Adaptive Systems	71
Complex Adaptive Financial Systems	146
Final Word	176
Definitions	178
Additional Readings	190
About the Author	191

Preface

Any system with many parts, pieces, elements, or agents that can choose and interact dynamically in nonlinear fashions would be considered complex Adaptive. These systems show highly nonlinear, cognitive, emergent, resilient, robust, and continuously adaptive behaviors in response to perturbations from their environments. They have a continuous formation of their temporal, spatial, or functional structures. They evolve through time, are dynamic, and can change their internal and external interactions or structures. It was also stated that Complex Adaptive Systems offer macro and holistic mental models to resolve current, challenging issues such as international socio-political tensions, urban development growth, electric grids, global internet networks, economic or financial dynamics, and technological innovations to name a few.

In classical mechanics, when we try to analyze the dynamics of a physical system, predictions mean the exact path of the system trajectory, and the equations of motion result in exact answers. The behavior of these systems is deterministic.

In quantum mechanics, the exact ability to simultaneously determine and predict an atomic size particle's position and momentum on an individual level is not possible. The behavior of these systems is probabilistic and not deterministic. There is uncertainty about the components and not about the way these components interact. We can still make deterministic predictions about the trajectory of a large set of particles collectively.

We have independent, changing components with nonlinear, dynamic, and adaptive interactions in Complex Adaptive Systems or CAS. In these systems, the components and their interactions are highly unpredictable. The predictability of the system trajectory and dynamics is extremely difficult (a lot more complicated than quantum mechanics) to understand. One can use simulations jointly with a stochastic approach for understanding these systems' holistic and general properties.

In this book, we start with the concept of Networks and Graphs. Networks are essential foundations for having a computational approach towards multi-elemental dynamic and nonlinear interactive systems. The computational approach towards Complex Adaptive Systems can be a potent and effective tool for understanding, measuring, and managing multi-agent dynamic behaviors.

The book is for individuals involved in the science of complexity and those curious to find ways to measure, compute, and manage Complex Adaptive Systems or CAS's behavior. From a practical and non-theoretical perspective, this is an introductory approach to quantify Complex Adaptive Systems' specific properties. Curious managers, instructors, engineering or MBA students, individuals, and practitioners that have been exposed to modeling concepts with a basic knowledge of statistics, calculus, differential equations, network, and system theory would enjoy this book.

At the end of each chapter, challenging questions and the corresponding suggested answers have been presented for discussion purposes. We should emphasize that the suggested answers should not limit the readers' creativity and innovative thought process to look for more comprehensible answers.

Introduction

It is difficult to determine if a system is simple, complicated, complex, and adaptive or not. We have gone through enough detail in describing the differences and various features that a change from simple to complex adaptive systems go through in our previous book on Understanding Complex Adaptive Systems[3].

Interconnected multi-agent systems will exhibit some degrees of complex and adaptive behavior. Depending on the degrees of agent-agent interconnections, cognition, and the nature of their connections (linear, nonlinear, deterministic, or statistical), they can exhibit lower or higher degrees of Complex Adaptive System (CAS) behavior. Examples of such system behaviors include swarm movements of bees, locus, birds, fishes, social unrests, a pandemic spreads such as Covid-19, and financial/economic instabilities due to specific events such as the 2008-09 2019 economic crisis.

It has been a significant challenge to understand the main factors contributing to Complex Adaptive behavior and the measurement, and management techniques required to control, contain, reverse or even, if needed, enhance their dynamics. We will try to refresh the comparative concepts concerning the input and output features for Simple (SS), Complicated (CS), Multi-Agent (MAS), Complex (COM), and Complex Adaptive Systems (CAS) presented before[4].

Using the definition for eight inputs and six outputs in the I/O measurement metrics, we assigned values and compared Simple (SS), Complicated (CS), Multi-Agent (MAS), Complex (COM), and Complex

---

[3] **Shayan, S. A. (2019)**,"Understanding Complex Adaptive Systems," Independent Publisher, Amazon.
[4] Ibid.

Adaptive Systems (CAS) from an input/output perspective. The results are shown in Table 1.

Table 1[5]
Comparing different systems
"Input / Output or I/O Measurement Metrics"

I/O Features	I/O	Code	Measurement Metrics				
			SS	CS	MAS	COM	CAS
Number of Agents or Nodes	I	NOA	1	2-5	2-5	2-5	2-5
Number of Positive Feedback and Feedforward loops	I	NPF	0	0-1	1-2	2-3	3-5
Number of Negative Feedback and Feedforward loops	I	NNF	0	0-1	1-2	2-3	3-5
Degree of Heterogeneity or Diversity of Agents	I	DHE	0	0-1	1-2	2-3	3-5
Degree of Mutuality/Agents Interconnections	I	DMI	0	0-1	1-2	2-3	3-5
Degree of Structural Flexibility or Degree of Freedom	I	DSF	0	0-1	1-2	2-4	3-5
Degree of Self-Similarity	I	DSS	0	0-1	1-2	1-2	1-5
Degree of Symmetry	I	DSY	4-5	1-2	1-3	2-4	3-5
Degree of Adaptation or Homeostasis	O	DAH	0	0	1-2	2-3	3-5
Degree of Contagion Effect	O	DCE	0	0	1-2	2-3	3-5
Degree of Self-Org., Emergence or Saltation	O	DES	0	0	1-2	2-3	3-5
Degree of Entropy or Information Uncertainty	O	DEI	0-1	0-2	2-3	2-4	3-5
Degree of Nonlinear Behavior	O	DNB	0	0	1-2	2-4	3-5

---

[5] Ibid, page 53, Table 2.

Degree of Resilience, Robustness, and Rigidity to Change	O	DRR	0	0-1	1-2	2-3	3-5
Degree of Complexity	O	DOC	0-1	1-2	2-3	2-4	3-5

Note1: None (0), Very Low (1), Low (2), Medium (3), High (4), and Very High (5)
Note 2: SS = Simple System, CS = Complicated System, MAS = Multi-Agent System, COS = Complex System, CAS = Complex Adaptive System

As shown in table1, at times, the differentiation between systems such as MAS and COM or COM and CAS is difficult. In such cases, we should look at the I/O features and deduce each system's behavioral nature.

We used the input/output dynamics to determine the relative degree of CAS behavior involved on a holistic and macro-level. Having more pieces, elements, nodes, agents, and structural similarities (structural self-similarity) do not necessarily cause a system to behave in a more complex adaptive manner. The agent's diversity, mutual interaction, connectivity, cooperation, collaboration, communication plus learning abilities or cognition (feedback/feedforward loops), and less structural flexibility or lower degrees of freedom (meaning more internal structure) are significant contributors to complex adaptive behavior. *More nonlinear interactions cause additional complexity, and more learning plus cognition causes increased adaptive behavior.*

In this book, we will try to be more quantitative and specific. We intend to follow and use a more quantitative approach to measure the Complex Adaptive features of a system.

All CAS structures include a simple or complicated network of interacting cognitive agents. For this reason, we will first explain what a network is and analyze it. We will also go over the concept of Graphs which becomes very important in understanding the interconnection and interactions among elements/agents in any network. Next, we will enhance the concept of networks and understand systems' concept, which is a more comprehensive analysis of how agent-agent interactions can behave and result in specific outputs. Using the concepts from the network theories, we will apply and use mathematical techniques to quantify Complex Adaptive Systems' behavior. Finally, we will apply all the covered concepts to the financial systems for practical purposes and see how we can benefit from a CAS viewpoint in such complex systems.

In the end, we have added a Definition section to clarifying certain additional terms. Throughout the book, we have presented several examples and posed many questions with suggested answers to develop the ideas presented.

Networks

A network is defined as a set of distinct objects, elements, agents, or nodes[6], with one or several connections, interactions or links, between them. The node-to-node links can be simple, one-sided, with constant strength, deterministic, static with linear effects, or be very complex two-sided, multiple node-to-node strengths with time-varying, self-interacting, non-deterministic nonlinear effects. Networks can have a simple or very complex web of layered structures and resulting dynamics.

In physical sciences, usually, force interactions are the same for all similar interacting bodies. In gases, liquids, and solids, all atoms or molecules behave and interact similarly. In network analysis, nodes or agents and their interactions can be different and specific, causing their mathematical analysis to be more complicated and difficult. In large networks, we are often forced to understand the network dynamics in terms of the statistical quantities of nodes and their interactions.

Network analysis is a required tool to understand complex systems. ***To understand networks, we should think as if everything is connected to everything else with varying degrees that can change through time.*** In other words, the strength and effect that each node creates on other nodes need to be understood through time. As an example, let us look at the following network structure:

---

[6] In network terms, we call each interacting body a node, but in later chapters on the complex adaptive system we will call them an agent.

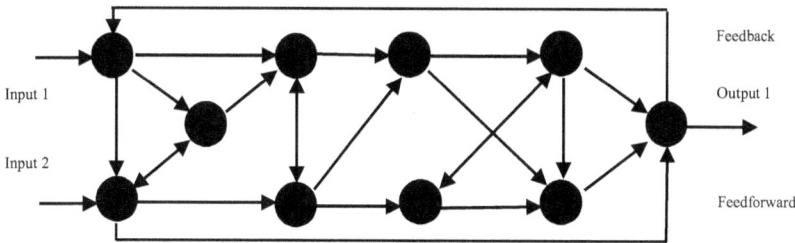

This network has ten nodes with two inputs, one output, one-sided and two-sided interacting nodes, one feedback, and one feedforward effect.

We can have sub-networks or network clusters within more extensive networks. By clusters, we mean a set of nodes at which each node's connections in the set mostly lead to other nodes in that set. As more node-to-node connections exist inside each set, a more specialized, complex, or localized cluster of nodes will exist. The node-to-node connections can have various strengths, positive or negative feedback[7], and feedforward[8] effects, leading to more complex hierarchical network

---

[7] **Positive or Negative Feedback Effects or Loops**. When we have dependent elements or agents and their outputs are added back as additional inputs, creating a part change or impact to the system's inputs, we observe a positive feedback loop, interaction, or recurrence effect. In such cases, we observe dependency of the nodes or agents on their outputs, which cause nonlinear effects to appear. The feedback effects can happen with delays (lags) and be positive (or additive) and reinforcing or negative (or subtractive) and dampening, diminishing, or stabilizing. Negative feedbacks usually absorb changes and settle the system into a state of equilibrium. Positive feedbacks could amplify changes, and after several loops or iterations can result in crash, instability, and failure of the system's outputs. Depending on the degree and weight of the feedback and after several iterations, nonlinearity in output is observed. The inputs and outputs in any system could be raw materials, energy, information, news, or any other elements defined by the system. Also see, **Shayan, S. A. (2019),**"Understanding Complex Adaptive Systems," page 131. Independent Publisher, Amazon.

[8] **Positive or Negative Feedforward Effects or Loops.** When the elements or agents are dependent, and their inputs are added forward as additional outputs creating a part change or impact to the system's outputs, we observe a feedforward loop, interaction, or reassurance effects. This internal interaction leads to nonlinearity effects in the systems output. The feedforward effects can happen with delays (lags) and be positive (or additive), amplifying, and reinforcing, leading to instability, crash, and system failure or negative (or subtractive), dampening, diminishing and stabilizing in effects on the outputs. Depending on the weight or degree of the feedforward, and after several loops and iterations, one can observe nonlinear behavior in the outputs. As in the Feedback

structures with more possibilities of dynamic network performances and complexities.

Networks are everywhere. Networks can interact with one another. An excellent article in the "Foreign Affairs" magazine[9] reviews the impacts of networks in our life. For more transparent understandings, we can classify networks as follows:

1. **Physical Networks (PN)**; have physical components or nodes (i.e., water, gas, oil pipelines, and electric grid distributions) with physical links between nodes, such as roads, pipelines, or fiber-optic networks. These are networks with actual physical nodes and links that have specific functions.
2. **Logical Networks (LN)**; have logical or thinking nodes (i.e., large connected parallel computer programs) with information flow as possible links. A map showing roads that connect cities is a logical network that replicates roads' actual physical network connecting cities.
3. **Fixed Networks (FN)**; have a fixed number of nodes and links. The network structure stays the same and does not change. It simplifies reality since we usually do not observe networks with nodes and links that stay constant forever.
4. **Information Networks (IN)**; have information points as their nodes (information flow through intranets, internet, and wireless telephones), and the links are through wifi, internet, fiber optics, radars, and satellites. These networks are a particular class on their own.
5. **Changing Networks (CN)**; have a varying number of nodes and links. These network structures change continuously, and the number of nodes and their links change through time and cause the network's function to behave dynamically and in a more complex fashion. Traffic network flows are examples of such changing networks. The number of cars and their routes change continuously through time and space.

---

effects, the inputs and outputs in the system could be raw materials, energy, information, news, or any other element defined by the system. Also see, **Shayan, S. A. (2019),** "Understanding Complex Adaptive Systems," page 132. Independent Publisher, Amazon.

[9] **Ferguson, N. (2017),** "The False Prophecy of Hyperconnection – How to Survive the Networked Age," Foreign Affairs Magazine September/October 2017, the Council of Foreign Relations, Inc.

6. **Hybrid Networks (HN);** changing physical networks, time-varying information networks, and the World Wide Web are examples of a changing or hybrid information network. Traffic flow in a big city is another example of a hybrid and physical network.
7. **Blended Networks (BN);** include any blend or combination of the other networks.

To quantify, measure, compare, and manage network performances, we have to determine and quantify important measurable attributes. Important network attributes include:

1. **Number of Nodes (NN);** more nodes can significantly impact the complexity of a network. How many nodes and the nature of their interactions impact the final output of a network.
2. **Node Homogeneity or Type (NH);** the more homogeneous the node types are, the easier it is to understand and model network behavior. Non-homogeneity or variations in node types adds to the output complexity and dynamics. One way to measure this feature is to determine the number of homogeneous node clusters in a network structure. The lower number of homogeneous node clusters can lead to more complex network structure and output dynamics.
3. **Number of Connections per Node (NCPN);** this is the total number of links nodes have with each other in the whole network. Higher links will affect the intensity of interactions and the complexity of node-to-node interrelationships. An average number of node links per node is a reasonable measure for this attribute. This attribute's higher value can increase network complexity and structure and help the network become more stable and face more dynamic network stability alternatives.
4. **The number of In Connections per Node (NICPN);** is the number of links with a node as their targets. The calculation is done for each node and on an average basis for the whole network.
5. **The number of Out Connections per Node (NOCPN);** is the number of links with a node as their source. The calculation is done for each node and on an average basis for the whole network.

6. **Node Coupling Strengths (NCS);** over and above the node to node links in a network and its existing node clusters, the *weight or coupling* strengths for each node to node link is essential. Not all links carry the same impact, weight, or coupling strength on other nodes. Some links have stronger coupling strength than others. In addition to the average value of coupling strengths per node, the distribution will be an essential measure for a better understanding of the nodes' impact map and the available options for network stability, resilience, and dynamic equilibrium.
7. **The number of Hamiltonian Cycles**[10] **(NHC);** if we start from a node and go through possible connections and can come back to the same node (which usually happens when there are feedback or feedforward connections) without having to go through a node twice, we have a Hamiltonian cycle in our network structure. We can have many Hamiltonian cycles in a given network or its underlying clusters. The more multiple and two-way connections there are among nodes, the more Hamiltonian cycles can be detected. More Hamiltonian cycles add to the complexity of the network and its dynamic behavior. We can use the number of Hamiltonian cycles in a network to understand how complex a network can potentially behave and where the sources of complexity are hidden in its structure. *One can even think about adding or eliminating node-to-node links to create or delete Hamiltonian cycles to increase or decrease the network's complexity.* One simple way to represent a Hamiltonian cycle is to show the nodes in each cycle, such as; [node1, node2, node3, node6, node8, node1]. Even a more accurate representation would be to show the out connection coupling strengths in brackets after each node in the series such as; [node1 (NCS1), node2 (NCS2), node3 (NCS3), node6 (NCS6), node8 (NCS8), node1 (NCS1)]. This representation is more transparent and shows at which connections there are increasing or decreasing coupling effects. The first and last nodes in every Hamiltonian cycle should be the same, and the coupling effects must be for links with a node as their source.

---

[10] **Hamiltonian Cycle;** A mathematical concept in Graph Theory.

8. **Average Network Entropy (ANE);** if each node in a network behaves stochastically and shows independent random behavior, its average randomness and behavior uncertainty can be calculated using Shannon's Entropy measure. If we have a network with n nodes, where node i have state j with the probability of observing node i in state j equal to p_j^i (the cumulative probability of node i in all j states will be equal to 1), then for each node, we can calculate the entropy[11] S_i as:

$$S_i = -\sum_{j=1}^{j=j} p_j^i \ln p_j^i$$

And for the network's average S_{NA}, we define;

$$S_{NA} = (-\sum_{i=1}^{i=n}\sum_{j=1}^{j=j} p_j^i \ln p_j^i)/n$$

For each node in this network, the summation of the probabilities of observing each state or p_j^i must add up to 1 or:

$$\sum_{j=1}^{j=j} p_j^i = 1$$

As the individual probability for node i or p_j^i gets larger and closer to 1, S_i gets closer to 0, which means with more certainty, we have less entropy per node. This reasoning is valid if the probability for all nodes i or p_j^i becomes larger and closer to 1, S_{NA} gets closer to 0, resulting in less entropy for the network. The reverse reasoning is also valid. If p_j^i becomes smaller and close to 0, S_{NA} gets larger, or for less certainty, we have more entropy. *Average Network Entropy*

---

[11] **Shannon's Entropy.** A concept used in information theory to measure and calculate the variability and uncertainty inherent in the information, outcome or behavior of a system. It was initially presented by Claude Shannon in 1948 as Information Entropy.

(ANE) explains how much variation, randomness, and uncertainty we should expect from a network's dynamic behavior. The higher the average network entropy is, the higher degrees of freedom its dynamic behavior will have. It also means the higher expected emergent and innovative behaviors we can expect.

9. **Node Dynamics (ND);** means understanding the network flow through the node to node links and determining the effects of coupling strengths between them through time. The node dynamics can be modeled, measured, and calculated using coupled nonlinear partial differential equations or more straightforward general statistical analysis techniques.

To understand network structures, the use of Graph Theory[12] and its applications have been very effective[13]. A network can be analyzed using

---

[12] **Graph Theory**; is a mathematical tool that facilitates the understanding and relationships between discrete objects. Leonard Euler a Swiss mathematician, astronomer, physicist, geographer and logician who by far is probably considered as one of the best and most creative mathematician in history, contributed to and used Graph theory to solve many practical and applied problems of his time (such as the famous Seven Bridges problem of Konigsberg and the topological relationship between **Vertex**, **E**dges and **F**aces of any polyhedral as **V- E + F = 2,** also known as the Euler's Polyhedron Formula). A graph consists of a set of vertices or nodes **(V)** and connections or links between these nodes called edges **(E)**. A graph **G** can be symbolized as **G = (V, E)**. Graphs are not diagrams but can be drawn to show their structures involved. The degree of vertices in a graph is the number of links coming in or going out of it. If an edge between two nodes is connected in one way a "directed" or non-symmetric edge exists and if it is in both ways a "undirected" or symmetric edge exists. In rare cases we have graphs called "completely connected" that have edges between every pair of edges or nodes. Some graphs are called "pseudo graphs" that have multiple edges and self-edges. The concept of self-edges for a node becomes very useful when we deal with Complex Adaptive Systems with cognitive nodes or agents. A walk with no repeated edges between vertices is called a "trail". A walk with no repeated vertices is called a "path". The shortest path between two vertices is called the "geodesic distance". If the endpoints of a trail are the same vertices (or we have a closed trail) then we have a "circuit" in our graph. We call a circuit with no repeated vertices a "cycle" in the graph. The vertices-vertices or node-node interactions of a graph can be shown through an "adjacency matrix" of zeros (for no vertices-vertices edges) and ones (for vertices-vertices edges). Graphs arise in many areas around us. Internet connections are an example of a very dynamic and interactive graph with users as nodes and internet connections as edges.

[13] **Simple network** model includes set of n nodes, elements or agents and connections between them with different weights. The dynamic relationship between the nodes can

Graph theory through a picture or graph showing all node-to-node connections and an interaction coupling strength matrix defining the nodal interconnections. For example, a network with ten nodes can be shown to have a potential of 10*10 node to node pictorial connections with varying weights attached to each connection. This structure can also be shown through a 10*10 interaction coupling strength matrix with the relative weights between every node in the corresponding rows and columns. This will be a complete network graph, including the *network picture* and the *corresponding interaction coupling matrix*. When networks interact and need to be compared or analyzed, network graphs and their interaction coupling strength matrix become very useful and meaningful. ***From a network point of view, complex adaptive systems are dynamic and evolving multilayer networks with time-varying different cognitive node-node interaction strengths and types.***

In general, a network with N nodes or agents *i*, will have two critical parameters. *First*, a node trigger parameter $T_i(t)$, that represents a node's ability to change inputs into outputs. This effect shows how an individual node output behaves in the network and can be represented as a scalar, vector, or even a tensor function that can change through time. The trigger parameter can show independent, random, or even cognitive behavior such as cells in the body or fish behavior in a swarm. *Second*, we define an interaction parameter $I_{ij}(t)$ to represent the interaction between nodes *i* and *j* through time. The value of this interaction parameter shows the strength of the given interaction between two nodes. The higher the value of this interaction parameter, the more rigid and more robust the node-to-node binding or coupling effects will be. The more interactions, the more adaptable, resilient, robust[14], and flexible the network stability will be.

There will be an optimum relationship between node interactions' strength and the number of interactions between nodes in any network. The more substantial node to node interaction means the less flexible, less adaptable, and more rigid the network structure will be. The lower the node to node interactions, the less and slower information dissipation and adaptability are observed. ***It is better to have more node-to-node***

---

be shown by a matrix of the order n by n. The matrix elements show the weights for each node-node interaction. This is called the Connection Matrix (see the Definition section).

[14] **Robustness;** is the ability of a system to correct errors in its structure, or get back to equilibrium when faced with shocks and perturbations.

connections with equally spread node-to-node interaction strengths (almost like a spider web) from a stability, adaptability, and resiliency perspective. For example, for solid materials (that have a very rigid and inflexible structure), the interaction parameter between atoms or molecules is relatively strong, for liquids less strong (with more flexible and adaptable structure), and gases the weakest (with very flexible and adaptable structure). The interaction parameter can be scalar (a number), vector (a number with direction), or a function (such as gravitational, electromagnetic, correlational, to name a few).

Therefore a network with N nodes or agents will have N node trigger parameters that can be represented by a vector with N elements of $T_1(t)$ to $T_N(t)$ and an interaction coupling strength matrix of N*N as follows:

$$T(t) = \begin{pmatrix} T_1(t) \\ \vdots \\ T_N(t) \end{pmatrix}$$

$$I(t) = \begin{pmatrix} I_{11}(t) & \cdots & I_{1N}(t) \\ \vdots & \ddots & \vdots \\ I_{N1}(t) & \cdots & I_{NN}(t) \end{pmatrix}$$

Any continuous change in the two defined network parameters through time can mathematically be represented as:

$$\frac{dT(t)}{dt} = \begin{pmatrix} \frac{dT_1}{dt} \\ \vdots \\ \frac{dt_N}{dt} \end{pmatrix}$$

$$\frac{dI(t)}{dt} = \begin{pmatrix} \frac{dI_{11}}{dt} & \cdots & \frac{dI_{1N}}{dt} \\ \vdots & \ddots & \vdots \\ \frac{dI_{N1}}{dt} & \cdots & \frac{dI_{NN}}{dt} \end{pmatrix}$$

The node trigger parameter vector and interaction coupling strength matrix can also be written for each sub-network or cluster θ separately as follows:

$$\boldsymbol{T^\theta}(t) = \begin{pmatrix} T_1^\theta(t) \\ \vdots \\ T_N^\theta(t) \end{pmatrix}$$

$$\boldsymbol{I^\theta}(t) = \begin{pmatrix} I_{11}^\theta(t) & \cdots & I_{1N}^\theta(t) \\ \vdots & \ddots & \vdots \\ I_{N1}^\theta(t) & \cdots & I_{NN}^\theta(t) \end{pmatrix}$$

Similarly, any continuous change in the two defined network parameters through time for each sub-network or cluster θ can mathematically be represented as:

$$\frac{d\boldsymbol{T^\theta}(t)}{dt} = \begin{pmatrix} \dfrac{dT_1^\theta}{dt} \\ \vdots \\ \dfrac{dT_N^\theta}{dt} \end{pmatrix}$$

$$\frac{d\boldsymbol{I^\theta}(t)}{dt} = \begin{pmatrix} \dfrac{dI_{11}^\theta}{dt} & \cdots & \dfrac{dI_{1N}^\theta}{dt} \\ \vdots & \ddots & \vdots \\ \dfrac{dI_{N1}^\theta}{dt} & \cdots & \dfrac{dI_{NN}^\theta}{dt} \end{pmatrix}$$

By defining and analyzing the node trigger parameters $T_i(t)$ or $T_i^\theta(t)$ and the interaction coupling strength matrix $\boldsymbol{I}(t)$ or $\boldsymbol{I^\theta}(t)$ one can understand the internal dynamics of a network and its components, sub-networks, or clusters. Due to the highly nonlinear nature of the node trigger parameters and the interaction coupling strength matrix, the linear algebra[15] technique's utilization is only possible when we make simplifying assumptions and linearize the node-node relationships. In doing so, we simplify node-node interacting features that are the sources of complex adaptive dynamic features, such as evolution, continuous or dynamic equilibrium, adaptiveness, and resiliency, that are of interest to us. In general, the use of linear algebra techniques in complex network analysis is not very useful.

---

[15] **Linear Algebra;** is a branch of mathematics that deals with finding Algebraic solutions to multiple variables in a system of linear equations. It utilizes concepts from vector and matrix algebra to solve systems of linear equations.

The interaction types between nodes or agents at times can be highly nonlinear. A more complicated network structure requires a more sophisticated mathematical tool such as tensors (we will not get into these more complicated structures in this book) for dynamic analysis.

In general, cluster θ with agent-agent interaction type λ will have trigger parameters defined as $T_i^{\lambda\theta}(t)$ and interaction coupling strengths defined as $I_{ij}^{\lambda\theta}(t)$, that can be scalar, vector, or a function.

In complex networks, each node's state or the trigger parameter functions and the interaction coupling strength matrix can mutually and dynamically affect one another and change through time. We can then observe a co-interacting, co-evolving behavior known as the adaptive phenomena between nodes and their respective interactions. Mathematically we can express this through a set of coupled nonlinear partial differential equations. The equation for each node can generally be defined as follows:

$$\frac{dT_i(t)}{dt} = F\,(I_{ij}(t), T_j(t))$$

$$\frac{dI_{ij}(t)}{dt} = K\,(I_{ij}(t), T_j(t))$$

Functions F and K can be deterministic, stochastic, nonlinear, or linear, which are dependent on the trigger parameters or state vectors $T_j(t))$ and interaction coupling strength matrix parameters $I_{ij}(t)$. For a sub-network or cluster θ we get the following general set of coupled partial differential equations:

$$\frac{dT_i^{\theta}(t)}{dt} = F\,(I_{ij}^{\theta}(t), T_j^{\theta}(t))$$

$$\frac{dI_{ij}^{\theta}(t)}{dt} = K\,(I_{ij}^{\theta}(t), T_j^{\theta}(t))$$

The first equation shows the dynamic change of the trigger parameter for node *i* through time regarding the interaction coupling strength matrix elements and the other node's trigger parameters. The second equation shows the interaction matrix's elements' dynamic change through time in terms of the interaction coupling strength matrix elements and trigger parameters for all nodes.

These are complicated equations to solve in closed mathematical forms. By using numerical or simulation techniques, it is possible to find approximate solutions for the way such highly co-interacting and co-evolving (adaptive) networks behave. The more nodes exist in a network, the more difficult the solutions are found. For large networks, heuristic and powerful simulation techniques have to be used to understand the network dynamics.

To better understand the forgone discussions, we will analyze three simple networks and examine their graphs, node trigger parameters, and interaction coupling strength matrix.

We will analyze three networks, A, B, and C, shown as follows:

Network A;

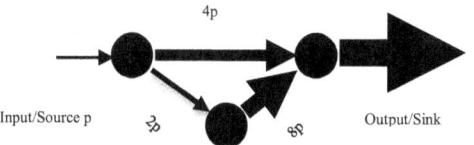

If each node is a water pump and amplifies its input flow pressure as described below:
- A water flow with pressure p into node1 (N1) acts as an input or a source in this network which turns into two outputs. One output amplifies the input pressure twice to 2p and the other four times to 4p.
- The output flow with twice the pressure is fed into another pump or node2 (N2), resulting in an output of four times or 8p (4x2p) the input pressure.
- The outputs of node1 and node2 are fed into node3 (N3) as inputs. If the final output pressure from node3 (N3) is a linear combination of the two input nodes, determine the node3 output or sink's flow pressure.

All node-node interactions create a final output equal to a function related to the pump-pump amplifications' linear cumulative effect. For the Network graph, classification, attributes, and the interaction coupling strength matrix, we have:

- **Network Graph:** Shown above
- **Network Classification:** Blended (Physical, Fixed)
- **Network Attributes:**
  - ✓ NN   3 nodes
  - ✓ NH   Non-homogeneous
  - ✓ NCPN   5 links with an avg. 1.7 per node including the source and sink
  - ✓ NICPN   4 links with an avg. 1.3 per node
  - ✓ NOCPN   4 links with an avg. 1.3 per node
  - ✓ NCS   Shown on the graph
  - ✓ NHC   None
  - ✓ ANE   0 with no variations in node behaviors
  - ✓ ND   input or source into N1 turns into two outputs, with one twice and the other four times the initial pressure. The output flow with twice the pressure feeds into another pump or N2, resulting in an output of four times the input pressure. N1 and N2 are fed into N3 as inputs that get added together and result in N3 output.

- **Input/Source:**   p
- **Output/Sink Result:** Node3 output pressure = 8p + 4p = 12p
- **Trigger Parameter Vector:**

$$T(t) = \begin{pmatrix} T_1(t) = 6p \\ T_2(t) = 8p \\ T_3(t) = 12p \end{pmatrix}$$

$$\frac{dT(t)}{dt} = \begin{pmatrix} 0 \\ 0 \\ 0 \end{pmatrix}$$

- **Interaction Coupling Strength Matrix:**

$$I(t) = \begin{pmatrix} 0 & 2p & 4p \\ 0 & 0 & 8p \\ 0 & 0 & 12p \end{pmatrix}$$

$$\frac{dI(t)}{dt} = \begin{pmatrix} 0 & 0 & 0 \\ 0 & 0 & 0 \\ 0 & 0 & 0 \end{pmatrix}$$

If the probability distribution of the total output pressure for each node is independently distributed and varies with time for each state a, as follows:

$$p(a) = e^{-at} \text{ and } \sum_{a=0}^{a=a} e^{-at} = 1$$

Find out the node trigger parameters, interaction coupling strength matrix, changes through time, and the Average Network Entropy or ANE. We get;

- **Trigger Parameter Vector:**

$$\boldsymbol{T}(t) = \begin{pmatrix} T_1(t) = 6e^{-at} \\ T_2(t) = 8e^{-at} \\ T_3(t) = 12e^{-at} \end{pmatrix}$$

$$\frac{d\boldsymbol{T}(t)}{dt} = \begin{pmatrix} -6ae^{-at} \\ -8ae^{-at} \\ -12ae^{-at} \end{pmatrix}$$

- **Interaction Coupling Strength Matrix:**

$$\boldsymbol{I}(t) = \begin{pmatrix} 0 & 2e^{-at} & 4e^{-at} \\ 0 & 0 & 8e^{-at} \\ 0 & 0 & 12e^{-at} \end{pmatrix}$$

$$\frac{d\boldsymbol{I}(t)}{dt} = \begin{pmatrix} 0 & -2ae^{-at} & -4ae^{-at} \\ 0 & 0 & -8ae^{-at} \\ 0 & 0 & -12ae^{-at} \end{pmatrix}$$

- **ANE equals to:**

$$S_1 = -\sum_{a=0}^{a=a} e^{-at}(-at)$$

$$S_2 = -\sum_{a=0}^{a=a} e^{-at}(-at)$$

$$S_3 = -\sum_{a=0}^{a=a} e^{-at}(-at)$$

$$ANE = S_{NA} = \frac{S_1 + S_2 + S_3}{3} = \sum_{a=0}^{a=a} ate^{-at}$$

If the probability distribution for each node depended on the distribution for other nodes, we had to also calculate and incorporate this dependency through more complicated coupled probability distributions, making our calculations very difficult. We will make the probability distributions' independent assumptions in most of our examples for simplicity and understanding. We have to use simulation models and more complicated mathematical calculations for the probability distributions' dependencies. We will not get into those calculations in this book.

Network B:

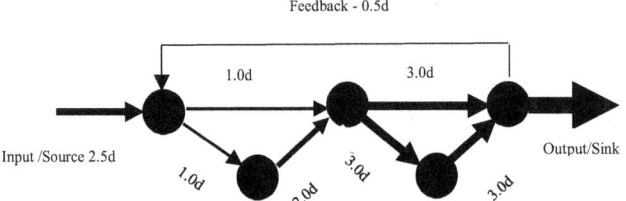

Each node is a computer that is connected, as the network graph shows. After proper processing, the data transfers are described as below:
- A data with the volume of 2.5 GB as input minus 0.5 GB (d equals to GB) as negative feedback effect (think of it as error corrections made from the final network output to the initial input data) is feed into node1 (N1). This data triggers specific calculations in node1 (N1) and turns the resulting data into two equal data volumes of 1.0 GB as outputs from this node.
- The first output 1.0 GB data from node1 (N1) is feed into node3 (N3), and the second is feed into node2 (N2). Data feed into node2 (N2) is doubled and is also feed into node3 (N3).
- The two data generated from node1 (N1) and node2 (N2) are added by node3 (N3) and generate two equal 3.0 GB data that each get feed into node4 (N4) and node5 (N5).
- The two data generated by node3 (N3) and node4 (N4) are added by node5 (N5) and generate one data source at which after 0.5 GB of data is fed back into node1 (N1) as correction

and negative feedback to the initial input data, the final output data is generated.
- Determine the final data output volume in Gigabyte (GB) terms for each node and node5 (N5) as output or sink.

To show the Network graph, classification, attributes, and the interaction coupling strength matrix, we have:

- **Network Graph:** Shown above
- **Network Classification:** Blended (Physical, Logical, Fixed)
- **Network Attributes:**
    - ✓ NN   5 nodes
    - ✓ NH   Non-homogeneous
    - ✓ NCPN   9 links with avg. 1.8 per node including the source and sink
    - ✓ NICPN   8 links with an avg. of 1.6 per node
    - ✓ NOCPN   8 links with an avg. of 1.6 per node
    - ✓ NCS   Shown on the graph
    - ✓ NHC   4 as follows:
        - ➢ N1(1.0d), N3(3.0d), N5(-0.5d), N1(1.0d)]
        - ➢ [N1(1.0d), N2(2.0d), N3(3.0d), N4(3.0d), N5(-0.5d), N1(1.0d)]
        - ➢ [N1(1.0d), N2(2.0d), N3(3.0d), N5(-0.5d), N1(1.0d)]
        - ➢ [N1(1.0d), N3(3.0d), N4(3.0d), N5(-0.5d), N1(1.0d)]
    - ✓ ANE   0 with no variations in avg. node behavior
    - ✓ ND   Input or source data is feed into N1 turns into two 1.0 GB equal outputs. One output is feed into N2 and gets doubled by N2 and gets added to the first output from N1 and feed into N3, and the total gets doubled by N3. The output from N3 is split into two equal 3.0 GB data sources, which are feed into N4 and N5. The output from N4 is the same as its input and gets added to the first output from N3 and feeds into N5. Output from N5 is the addition of the two inputs from N3 and N4. A data source error correction of 0.5 GB is made to the final network output from N5 and negatively affects the initial network data source. Through this process, the initial data is amplified, and its errors are corrected through several loops.

- **Initial Input/Source:**   2.5 GB
- **First Output/Sink Result:**   N5 = 3.0 + 3.0 - 0.5 = 5.5 GB
- **Trigger Parameter Vector:**

$$T(t) = \begin{pmatrix} T_1(t) = 2.0d \\ T_2(t) = 2.0d \\ T_3(t) = 6.0d \\ T_4(t) = 3.0d \\ T_5(t) = Output(t) = 3.0d + 3.0d - 0.5d = 5.5d \end{pmatrix}$$

$$\frac{d\mathbf{T}(t)}{dt} = \begin{pmatrix} 0 \\ 0 \\ 0 \\ 0 \\ 0 \end{pmatrix}$$

- **Interaction Coupling Strength Matrix:**

$$\mathbf{I}(t) = \begin{pmatrix} 0 & 1d & 1d & 0 & 0 \\ 0 & 0 & 2d & 0 & 0 \\ 0 & 0 & 0 & 3d & 3d \\ 0 & 0 & 0 & 0 & 3d \\ -0.5d & 0 & 0 & 0 & 6d \end{pmatrix}$$

$$\frac{d\mathbf{I}(t)}{dt} = \begin{pmatrix} 0 & 0 & 0 & 0 & 0 \\ 0 & 0 & 0 & 0 & 0 \\ 0 & 0 & 0 & 0 & 0 \\ 0 & 0 & 0 & 0 & 0 \\ 0 & 0 & 0 & 0 & 0 \end{pmatrix}$$

If the probability distribution of output data for each node is independently distributed and varies with time for each state d, as follows:

$$P(d) = dt^{-1} \text{ and } \sum_{d=0}^{d=d} dt^{-1} = 1$$

Find out the node Trigger Parameters, Interaction Parameter Strength Matrix, their changes through time, and the Average Network Entropy or ANE. We get;

- **Trigger Parameter Vector:**

$$\mathbf{T}(t) = \begin{pmatrix} T_1(t) = 2dt^{-1} \\ T_2(t) = 2dt^{-1} \\ T_3(t) = 6dt^{-1} \\ T_4(t) = 3dt^{-1} \\ T_5(t) = 5.5dt^{-1} \end{pmatrix}$$

$$\frac{d\boldsymbol{T}(t)}{dt} = \begin{pmatrix} -2dt^{-2} \\ -2dt^{-2} \\ -6dt^{-2} \\ -3dt^{-2} \\ -5.5dt^{-2} \end{pmatrix}$$

- **Interaction Coupling Strength Matrix:**

$$\boldsymbol{I}(t) = \begin{pmatrix} 0 & dt^{-1} & dt^{-1} & 0 & 0 \\ 0 & 0 & 2dt^{-1} & 0 & 0 \\ 0 & 0 & 0 & 3dt^{-1} & 3dt^{-1} \\ 0 & 0 & 0 & 0 & 3dt^{-1} \\ -0.5dt^{-1} & 0 & 0 & 0 & 6dt^{-1} \end{pmatrix}$$

$$\frac{d\boldsymbol{I}(t)}{dt} = \begin{pmatrix} 0 & -dt^{-2} & -dt^{-2} & 0 & 0 \\ 0 & 0 & -2dt^{-2} & 0 & 0 \\ 0 & 0 & 0 & -3dt^{-2} & -3dt^{-2} \\ 0 & 0 & 0 & 0 & -3dt^{-2} \\ 0.5dt^{-2} & 0 & 0 & 0 & -6dt^{-2} \end{pmatrix}$$

- **ANE equals to:**

$$S_1 = -\sum_{d=0}^{d=d} dt^{-1} (\ln d - \ln t)$$

$$S_2 = -\sum_{d=0}^{d=d} dt^{-1} (\ln d - \ln t)$$

$$S_3 = -\sum_{d=0}^{d=d} dt^{-1} (\ln d - \ln t)$$

$$S_4 = -\sum_{d=0}^{d=d} dt^{-1} (\ln d - \ln t)$$

$$S_5 = -\sum_{d=0}^{d=d} dt^{-1} (\ln d - \ln t)$$

$$ANE = \frac{S_1 + S_2 + S_3 + S_4 + S_5}{5} = \sum_{d=0}^{d=d} dt^{-1} (\ln t - \ln d)$$

Network C:

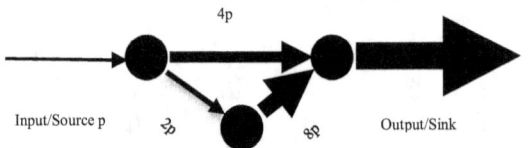

Input/Source p   4p   8p   Output/Sink

If with network A, the output pressure varies statistically, following a continuous and independent normal distribution function as follows:

$$P(p) = \alpha e^{-0.5\left(\frac{(p-\mu)^2}{\sigma^2}\right)} \quad \text{and} \quad \sum_{p=0}^{p=p} \alpha e^{-0.5\left(\frac{(p-\mu)^2}{\sigma^2}\right)} = 1$$

Where μ is the mean pressure, σ is the standard deviation of the pressure and α is a parameter dependent on the inverse of standard deviation σ. For the sake of simplicity, we assume that μ, σ, and α are equal for all nodes and are constant throughout the network. Also, each node's distribution is independent of the others (in reality, these parameters do change, and nodes are dependent on other nodes, which makes the analysis more complicated)[16].

---

[16] **Combining Normal Probability Distributions;** The combined resulting effects of adding, subtracting or multiplying n independent (with no correlations) normally distributed random variables with μ_i as the mean and σ_i as the standard deviation are shown below (we leave the proofs for these results and in case of existing correlations between the variables, for the interested readers to pursue);

- If adding or subtracting n independent normally distributed random variables,

$\mu_T = \sum_1^n \mu_i$ Expected Mean (for subtraction put – sign for each *i*)
$\sigma_T^2 = \sum_1^n \sigma_i^2$ Expected Variance (the same for subtraction)

For only two variable 1 and 2 we get;

Find out the node Trigger Parameters, Interaction Parameter Strength Matrix, their changes through time, and the Average Network Entropy or ANE, in addition to the expected mean and variance[17] of the outputs at each node. We get;

- **Trigger Parameter Vector:**

$$T(t) = \begin{pmatrix} T_1(t) = 6P(p) = 6\alpha e^{-0.5\left(\frac{p-\mu}{\sigma}\right)^2} \\ T_2(t) = 8P(p) = 8\alpha e^{-0.5\left(\frac{p-\mu}{\sigma}\right)^2} \\ T_3(t) = 12P(p) = 12\alpha e^{-0.5\left(\frac{p-\mu}{\sigma}\right)^2} \end{pmatrix}$$

$\mu_T = \mu_1 + \mu_2$ Expected Mean (for subtraction put – sign for each i)
$\sigma_T^2 = \sigma_1^2 + \sigma_2^2$ Expected Variance (the same for subtraction)

- If multiplying we get,

$\mu_T = \prod_1^n \mu_i$ Expected Mean
$\sigma_T^2 = \prod_1^n (\sigma_i^2 + \mu_i^2) - \prod_1^n \mu_i^2$ Expected Variance

For only two variable 1 and 2 we get;

$\mu_T = \mu_1 \mu_2$ Expected Mean
$\sigma_T^2 = (\sigma_1^2 + \mu_1^2)(\sigma_2^2 + \mu_2^2) - \mu_1^2 \mu_2^2$ Expected Variance

And if an independent normally distributed random variable is multiplied by a constant c;

$\mu_T = c\mu_i$ Expected Mean
$\sigma_T^2 = c^2 \sigma_i^2$ Expected Variance

- In general, if n independent normally distributed random variables are multiplied by constants c_i and then added linearly we get;

$X_T = \sum_1^n c_i X_i$ (Linear combination of independent random variables)
$\mu_T = \sum_1^n c_i \mu_i$ Expected Mean
$\sigma_T^2 = \sum_1^n c_i^2 \sigma_i^2$ Expected Variance

Note: The sign $\sum$ means summing up and $\prod$ means multiplying all terms i from 1 to n.

---

[17] **Variance;** is the square of the standard deviation σ or σ^2.

$$\frac{d\mathbf{T}(t)}{dt} = \begin{pmatrix} 6\frac{dP(p)}{dt} = 0 \\ 8\frac{dp(p)}{dt} = 0 \\ 12\frac{dp(p)}{dt} = 0 \end{pmatrix}$$

- **Interaction Coupling Strength Matrix:**

$$I(t) = \begin{pmatrix} 0 & 2\alpha e^{-0.5\left(\frac{(p-\mu)^2}{\sigma^2}\right)} & 4\alpha e^{-0.5\left(\frac{(p-\mu)^2}{\sigma^2}\right)} \\ 0 & 0 & 8\alpha e^{-0.5\left(\frac{(p-\mu)^2}{\sigma^2}\right)} \\ 0 & 0 & 12\alpha e^{-0.5\left(\frac{(p-\mu)^2}{\sigma^2}\right)} \end{pmatrix}$$

$$\frac{d\mathbf{I}(t)}{dt} = \begin{pmatrix} 0 & 0 & 0 \\ 0 & 0 & 0 \\ 0 & 0 & 0 \end{pmatrix}$$

- **Expected mean and variance of the outputs at each node (use the information in footnote 14):**

  ✓ **For node1 output to node3;**
  $\mu_{T13} = 4\mu$ Expected Mean
  $\sigma^2_{T13} = 16\sigma^2$ Expected Variance

  ✓ **For node1 output to node2;**
  $\mu_{T12} = 2\mu$ Expected Mean
  $\sigma^2_{T12} = 4\sigma^2$ Expected Variance

  ✓ **For node2 output to node3;**
  $\mu_{T23} = 8\mu$ Expected Mean
  $\sigma^2_{T23} = 64\sigma^2$ Expected Variance

  ✓ **For node3 output;**
  $\mu_{T3} = 4\mu + 8\mu = 12\mu$ Expected Mean
  $\sigma^2_{T3} = 16\sigma^2 + 64\sigma^2 = 80\sigma^2$ Expected Variance

- **ANE equals to:**

$$S_1 = -\sum_{p=0}^{p=p} \alpha e^{-0.5\left(\frac{p-\mu}{\sigma}\right)^2} \left(\ln \alpha - 0.5\left(\frac{p-\mu}{\sigma}\right)^2\right)$$

$$S_2 = -\sum_{p=0}^{p=p} \alpha e^{-0.5\left(\frac{p-\mu}{\sigma}\right)^2} \left(\ln \alpha - 0.5 \left(\frac{p-\mu}{\sigma}\right)^2\right)$$

$$S_3 = -\sum_{p=0}^{p=p} \alpha e^{-0.5\left(\frac{p-\mu}{\sigma}\right)^2} \left(\ln \alpha - 0.5 \left(\frac{p-\mu}{\sigma}\right)^2\right)$$

$$ANE = \frac{S_1 + S_2 + S_3}{3} = \sum_{p=0}^{p=p} \alpha e^{-0.5\left(\frac{p-\mu}{\sigma}\right)^2} \left(0.5 \left(\frac{p-\mu}{\sigma}\right)^2 - \ln \alpha\right)$$

As it can be imagined, if we had to deal with coupled or dependent probability distributions for each node, our computations would have been extremely cumbersome, and computer simulations had to be utilized. We will leave these more complicated calculations to the interested readers to follow.

A network structure can be impacted, influenced, or triggered through the following three primary sources:

1. **Inside Sources (IS);** these are sources from within the network structure. They are self-induced effects created by one or several correlated or non-correlated nodes. These effects can be *continuous, pulsed, irregular, or stochastic.* Cognitive nodes or agents have this kind of self-inducing effect in a network.
2. **Outside Sources (OS);** these are sources from outside of the network structure that can influence more than one node in a structure. These sources can be *continuous, pulsed, irregular, or stochastic.* The network responds when such effects are trickled down through the affected nodes.
3. **Hybrid Sources (HS);** are a combination of both the outside and inside sources. It happens when an outside source plus inside self-induced sources are created either in a random non-correlated or correlated manner (when you have a correlated manner, the contagion effect can be observed). It usually starts with an outside source causing a lagged inside source as a response that gets trickled down through the network in a complex, co-evolving manner. It is how our brain, a complex network of interacting neurons, behaves.

Think of how the brain of a soccer player that is exposed to many stochastic impulses in a game functions. It is a perfect example of having been exposed to IS and OS-like sources.

It is evident that in any network behaving statistically (which most networks do), the more nodes or agents we have in combination with more significant node-to-node interactions with varying coupling effects through time, the more complex and diverse the network behavior will emerge. The various statistical behaviors of inputs and trigger parameters can lead to diverse and complex network output and dynamics. The probabilistic features of networks trigger the concept of emergence and innovation in network dynamics. It is an instrumental concept in Complex Adaptive System analysis. We will get back to this concept in the later chapters.

As the last note in this section, when a node or agent's output cannot be determined or analyzed through a simple linear combination (adding or subtracting) of the inputs, we have a nonlinear dynamic behavior in our system. This means that the superposition principle does not work, or the whole is not the sum of its parts. All complex systems have nonlinear behavior, and their resulting outputs cannot be determined from a linear combination of their inputs. Nonlinear dynamics can become unpredictable and unstable under certain conditions, a unique subject known as nonlinear dynamics and Chaos (see Definitions).

Questions

Question 1. Let us assume that we have the following network structure:

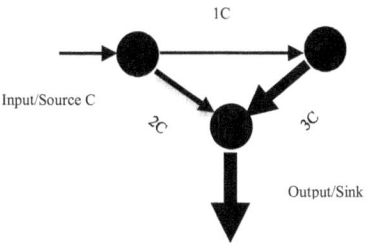

If the first and second nodes are biotech reactors connected in series that amplify their interferon input concentration C as shown on the graph and the last node is a mixer that linearly adds the two product concentrations into the final output or sink, find the network classification, attributes, and the interaction coupling strength matrix for this network.

Answer 1:
We have,

- **Network Graph:** Shown above
- **Network Classification:** Blended (Physical, Fixed)
- **Network Attributes:**
  - ✓ NN   3 nodes
  - ✓ NH   Non-homogeneous
  - ✓ NCPN   5 links with an avg. 1.7 per node including the source and sink
  - ✓ NICPN   4 links with an avg. 1.3 per node
  - ✓ NOCPN   4 links with an avg. 1.3 per node
  - ✓ NCS   Shown on the graph
  - ✓ NHC   None
  - ✓ ANE   0 with no variations in avg. node behavior
  - ✓ ND   input or source into N1 turns into two outputs with one twice and the other equal to the initial concentration. The output equaling the initial concentration feeds into another reactor, N2, which results in an output with three times the input concentration. The outputs of N1 and N2 are added together and result in N3 output.

- **Input/Source:**   C
- **Output/Sink Result:** Node3 output concentration 2C + 3C = 5C

- **Trigger Parameter Vector:**

$$T(t) = \begin{pmatrix} T_1(t) = C + 2C = 3C \\ T_2(t) = 3C \\ T_3(t) = 2C + 3C = 5C \end{pmatrix}$$

$$\frac{dT(t)}{dt} = \begin{pmatrix} 0 \\ 0 \\ 0 \end{pmatrix}$$

- **Interaction Coupling Strength Matrix:**

$$I(t) = \begin{pmatrix} 0 & 1C & 2C \\ 0 & 0 & 3C \\ 0 & 0 & 5C \end{pmatrix}$$

$$\frac{dI(t)}{dt} = \begin{pmatrix} 0 & 0 & 0 \\ 0 & 0 & 0 \\ 0 & 0 & 0 \end{pmatrix}$$

Question 2. If in the previous question, the probability distribution of concentration for the output source for each node (biotech reactor) was independent and normally distributed as follows:

$$P(C) = C_0 e^{-0.5\left(\frac{(C-\mu)^2}{\sigma^2}\right)} \quad \text{and} \quad \sum_{C=0}^{C=C} C_0 e^{-0.5\left(\frac{(C-\mu)^2}{\sigma^2}\right)} = 1$$

With an equal average of μ and standard deviation of σ for all nodes, find the network entropy for each node and the average Network Entropy or ANE.

Answer 2:
We have,

$$S_1 = -\sum_{C=0}^{C=C} C_0 e^{-0.5\left(\frac{C-\mu}{\sigma}\right)^2} \left(\ln C_0 - 0.5\left(\frac{C-\mu}{\sigma}\right)^2\right)$$

$$S_2 = -\sum_{C=0}^{C=C} C_0 e^{-0.5\left(\frac{C-\mu}{\sigma}\right)^2} \left(\ln C_0 - 0.5\left(\frac{C-\mu}{\sigma}\right)^2\right)$$

$$S_3 = -\sum_{C=0}^{C=C} C_0 e^{-0.5\left(\frac{C-\mu}{\sigma}\right)^2} \left(\ln C_0 - 0.5\left(\frac{C-\mu}{\sigma}\right)^2\right)$$

$$ANE = \frac{S_1 + S_2 + S_3}{3} = \sum_{C=0}^{C=C} C_0 e^{-0.5\left(\frac{C-\mu}{\sigma}\right)^2} \left(0.5\left(\frac{C-\mu}{\sigma}\right)^2 - \ln C_0\right)$$

Question 3. If in the previous question, the probability distribution of concentration for the output source for each node (biotech reactor) followed a different independent distribution as follows:

$$P(C) = C_0 e^{-t} \quad \text{and} \quad \sum_{C=0}^{C=C} C_0 e^{-t} = 1$$

Find the network entropy for each node and the average Network Entropy or ANE.

Answer 3:
We have,

$$S_1 = -\sum_{C=0}^{C=C} C_0 e^{-t} (\ln C_0 - t)$$

$$S_2 = -\sum_{C=0}^{C=C} C_0 e^{-t} (\ln C_0 - t)$$

$$S_3 = -\sum_{C=0}^{C=C} C_0 e^{-t} (\ln C_0 - t)$$

$$ANE = \frac{S_1 + S_2 + S_3}{3} = \sum_{C=0}^{C=C} C_0 e^{-t} (t - \ln C_0)$$

Question 4. Find the network entropy for each node, and the average Network Entropy or ANE as t becomes very large in the previous question.

Answer 4:
As time t becomes very large, we have,

$$S_1 = -\sum_{C=0}^{C=C} C_0 e^{-t} (\ln C_0 - t) = 0$$

$$S_2 = -\sum_{C=0}^{C=C} C_0 e^{-t} (\ln C_0 - t) = 0$$

$$S_3 = -\sum_{C=0}^{C=C} C_0 e^{-t} (\ln C_0 - t) = 0$$

$$ANE = \frac{S_1 + S_2 + S_3}{3} = 0$$

As time goes on, the system becomes symmetric or stable, and its entropy moves toward zero.

Question 5. Let us assume that we have the following network structure;

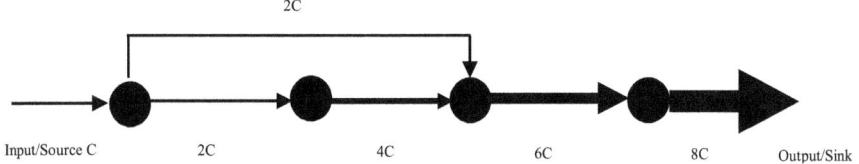

If all nodes are different chemical reactors connected in series that increase the concentration C of a given product into the final output or sink, find network classification, attributes, and the interaction coupling strength matrix for this network.

Answer 5:
We have,

- **Network Graph:** Shown above
- **Network Classification:** Blended (Physical, Fixed)
- **Network Attributes:**
  - ✓ NN   4 nodes
  - ✓ NH   Non-homogeneous
  - ✓ NCPN   6 links with an avg. 1.5 per node including the source and sink
  - ✓ NICPN   5 links with an avg. 1.25 per node
  - ✓ NOCPN   5 links with an avg. 1.25 per node

- ✓ NCS   Shown on the graph
- ✓ NHC   None
- ✓ ANE   0 with no variations in avg. node behavior
- ✓ ND   input or source concentration C is feed into node N1, which turns into two outputs with 2C each, where one feeds into node N2 and the other into node N3. The feed into Node N2 results in output 4C, where it is feed into node N3 in conjunction with the 2C feed from node N1. The output from N3 is 6C and is feed into node N4, which results in an output or sink of 8C.

- **Input/Source:** C
- **Output/Sink Result:** Node4 output concentration = 8C
- **Trigger Parameter Vector:**

$$T(t) = \begin{pmatrix} T_1(t) = 4C \\ T_2(t) = 4C \\ T_3(t) = 6C \\ T_4(t) = 8C \end{pmatrix}$$

$$\frac{dT(t)}{dt} = \begin{pmatrix} 0 \\ 0 \\ 0 \\ 0 \end{pmatrix}$$

- **Interaction Coupling Strength Matrix:**

$$I(t) = \begin{pmatrix} 0 & 2C & 2C & 0 \\ 0 & 0 & 4C & 0 \\ 0 & 0 & 0 & 6C \\ 0 & 0 & 0 & 8C \end{pmatrix}$$

$$\frac{dI(t)}{dt} = \begin{pmatrix} 0 & 0 & 0 & 0 \\ 0 & 0 & 0 & 0 \\ 0 & 0 & 0 & 0 \\ 0 & 0 & 0 & 0 \end{pmatrix}$$

Question 6. In the previous question, the concentration probability distribution of the output source for each node (chemical reactor) is independent and has the following mathematical distribution:

$$P(C) = C_0 e^{-C\, C_0} \text{ and } \sum_{C=0}^{C=C} C_0 e^{-C\, C_0} = 1$$

We have a negative exponential distribution. Find the network entropy for each node and the average Network Entropy or ANE.

Answer 6:
We have,

$$S_1 = -\sum_{C=0}^{C=C} C_0 e^{-C C_0} (\ln C_0 - (CC_0))$$

$$S_2 = -\sum_{C=0}^{C=C} C_0 e^{-C C_0} (\ln C_0 - (CC_0))$$

$$S_3 = -\sum_{C=0}^{C=C} C_0 e^{-C C_0} (\ln C_0 - (CC_0))$$

$$S_4 = -\sum_{C=0}^{C=C} C_0 e^{-C C_0} (\ln C_0 - (CC_0))$$

$$ANE = \frac{S_1 + S_2 + S_3 + S_4}{4} = \sum_{C=0}^{C=C} C_0 e^{-CC_0} ((CC_0) - \ln C_0)$$

Question 7. If in the previous question, $C_0 = 1 \, g/mole$, what would happen to the network entropy for each node and the average Network Entropy.

Answer 7:
We get,

$$S_1 = \sum_{C=0}^{C=C} C e^{-C}$$

$$S_2 = \sum_{C=0}^{C=C} C e^{-C}$$

$$S_3 = \sum_{C=0}^{C=C} C e^{-C}$$

$$S_4 = \sum_{C=0}^{C=C} Ce^{-C}$$

$$ANE = \frac{S_1 + S_2 + S_3 + S_4}{4} = \sum_{C=0}^{C=C} Ce^{-C}$$

As C increases, individual node and network average entropies decrease.

Question 8. In a network, as the Number of Nodes (NN) and their connections increase, what happens to the Number of Hamiltonian Cycles (NHC) and Average Network Entropy (ANE).

Answer 8:
As the NNs increase with their corresponding connections, ANE will increase. This is because the possibility of variations increases (this can be seen through the entropy equation). On the other hand, NHC will increase if node connections change through additional direct and indirect (multiple nodes) feedback and feedforward loops created in the network.

Question 9. In an extensive network with n independent nodes, half of the nodes have the same homogeneity and independent probability distribution for parameter x as follows:

$$P(x) = X_0 e^{-0.5\left(\frac{X-\mu}{\sigma}\right)^2} \quad and \quad \sum_{X=0}^{X=X} X_0 e^{-0.5\left(\frac{X-\mu}{\sigma}\right)^2} = 1$$

And the other half as:

$$P(x) = xt^{-1} \quad and \quad \sum_{X=0}^{X=X} xt^{-1} = 1$$

In here t is time. Find out what is the Average Network Entropy when time is considerable (goes to infinity).

Answer 9:
We have:

$$S_{\frac{n}{2}} \text{ (1st half)} = -\left(\frac{n}{2}\right) \cdot \sum_{X=0}^{X=X} X_0 e^{-0.5\left(\frac{X-\mu}{\sigma}\right)^2} \left(\ln X_0 - 0.5 \left(\frac{X-\mu}{\sigma}\right)^2\right)$$

$$S_{\frac{n}{2}} \text{ (2nd half)} = -\left(\frac{n}{2}\right) \cdot \sum_{X=0}^{X=X} xt^{-1} (\ln x - \ln t)$$

$$ANE = \frac{-\left(\frac{n}{2}\right) \cdot \sum_{X=0}^{X=X} X_0 e^{-0.5\left(\frac{X-\mu}{\sigma}\right)^2} \left(\ln X_0 - 0.5 \left(\frac{X-\mu}{\sigma}\right)^2\right) + xt^{-1}(\ln x - \ln t)}{n}$$

$$ANE = \frac{-\sum_{X=0}^{X=X} X_0 e^{-0.5\left(\frac{X-\mu}{\sigma}\right)^2} \left(\ln X_0 - 0.5 \left(\frac{X-\mu}{\sigma}\right)^2\right) + xt^{-1}(\ln x - \ln t)}{2}$$

If t becomes large, ANE becomes:

$$ANE = \frac{-\sum_{X=0}^{X=X} X_0 e^{-0.5\left(\frac{X-\mu}{\sigma}\right)^2} \left(\ln X_0 - 0.5 \left(\frac{X-\mu}{\sigma}\right)^2\right)}{2}$$

As time passes by, ANE becomes equal to the ANE of the first half of nodes because the second half nodes' entropy goes to zero. The second half nodes reach a stable behavior at large t, and the only source of entropy remaining in the network will be caused by the variations of x in the first half nodes.

Question 10. In question 9, under which condition would the ANE of the first half nodes become zero when t is large.

Answer 10:
When the standard deviation or σ of all the first half nodes are equal to zero, this only happens when X becomes fixed or constant. In this case, the network will have fixed or non-statistical dynamics with no entropy.

Question 11. Explain the positive and negative effects of having highly interconnected vs. less connected social networks.

Answer 11:
The comparison can be shown in the table below:

Effects of Social Network Interconnections		
Network Interconnections	Positives	Negatives
Low	Stable and Non-volatile behavior.	Information gap.
High	Increase of fast, uncensored, accessible voice, image, and idea exchange flows. Ability to detect criminal acts.	Misinformation can cause panic, financial, legal, and political fraud. Can lead to prejudice, authoritarianism, isolationism, and nationalism at a global level.

Question 12. It has been stated, that globally we have six or fewer social connections (also known as degrees) away from each other. What will happen if the 6 degrees of separation between each individual and any other person on the planet increases to 8? What happens if it is reduced to 3 (it has been claimed that the average degree of separation for Facebook users is 3.57)? Look at this from a Netizen's perspective (active citizens connected through networks such as the internet or other social media).

Answer 12:
The comparison can be shown in the table below:

Effects of Different Degrees of Separation between individuals	
Degrees of Separation	Results in
8.00	Information gap increases. We have reduced the ability to interact and communicate on a global level. More ability to censor information. Less able to detect fraudulent acts.
3.00	The global flow of uncensored, accessible voice, image, and ideas will increase. Better ability to detect criminal acts. Increase misinformation can cause increased panic, financial, legal, and political fraud. Can increase more prejudice, authoritarianism, isolationism, and nationalism at a global level.

Question 13. Large social networks have many nodes with a vast number of edges. These networks have many more nodes with multiple edges as compared to randomly generated networks. Social networks are dynamic, expand, and their nodes gain new edges as they evolve and grow. Compare the two in a simple table.

Answer 13:
The comparison can be shown in the table below:

\multicolumn{2}{c}{Comparing Large Social and Randomly Generated Networks}	
Type of Network	**General Features**
Randomly Generated	Lower ability to have live, uncensored information where it is needed. Less ability to interact, create networks and communicate on a global level. More ability to censor information.
Large Social	High ability to exchange unregulated, uncensored, accessible voice, image, and ideas globally. High ability to detect criminal acts. High ability of misinformation causing increased panic, financial, legal, and political fraud. High ability to create prejudice, authoritarianism, isolationism, and nationalism at the global level. Increase in innovative and emergent results and structures in such networks.

Question 14. If we accept that large networks are complex adaptive systems that constantly change their connections or edges, what would be their significant features compared to large non-adaptive networks?

Answer 14:
Large networks that are considered complex adaptive systems will have cognitive abilities at their node or agent levels. Cognition can lead to continuous stability, adjustments, and equilibrium resulting from external shocks. They will have more resilient and innovative dynamics. The more nodes or agents have cognitive features, the more adaptable, resilient, adjustable, and innovative the network dynamics.

Question 15. The danger with large social networks is the control of the network in the hands of a few. How could this control structure be considered a threat?

Answer 15:
The central control of large social networks regarding network traffic, information flow, and behavioral patterns of agents can potentially lead to fraud, misuse, and concentration of power and information in the hands of a few. It is potentially a precedent to authoritarian manipulation and unfair treatment of nodes, agents, and society at large. It is a potential problem with large network setups with central or concentrated controls (having one or several nodes connected to all other nodes through multiple edges). One way to resolve this potential problem is through regulation by the respective governments. Due to the global nature of large social networks, coordinated government regulations could be a challenging task. Better yet, an international self-regulatory mechanism by the network agents, very similar to the approach used in the crypto-currency markets using the Blockchain concept, can become more effective.

Systems

In this chapter, we will assume that the readers are familiar with the basic concepts of systems. We will cover the basics of systems, emphasizing coupled, multi-variable, nonlinear, continuous, or discrete systems as much as needed for preparation to understand Complex Adaptive Systems in the next chapter.

Systems are more structured versions of multi-nodal or multi-agent networks. For every system, we have one or several inputs and one or several outputs. Inputs create changes or pulses through one or several of the nodes or agents, the effects at which would go through and an interacting process (sometimes called algorithms) resulting in one or several outputs. All systems have boundaries that allow various degrees of interaction with their outside environments. **Open Systems** are porous and allow ongoing interactions or communications with their environments. They are more dynamic and complicated to understand and manage. **Semi-Open Systems** are semi-porous, have limited or partial interactions with their environments, and are better understood and managed. **Closed systems** are non-porous and have no interactions or communications with their environments. They do not get any influences from the outside, making them easier to understand and manage. Systems can have simple linear deterministic input-process-output structures or complex non-deterministic, highly interacting nonlinear, multiple layers of feedback/feedforward structures[18]. The pictures below show how system structures and flows are defined.

---

[18] Also see, **Shayan, S. A. (2019)**,"Understanding Complex Adaptive Systems," Independent Publisher, Amazon.

Open Systems

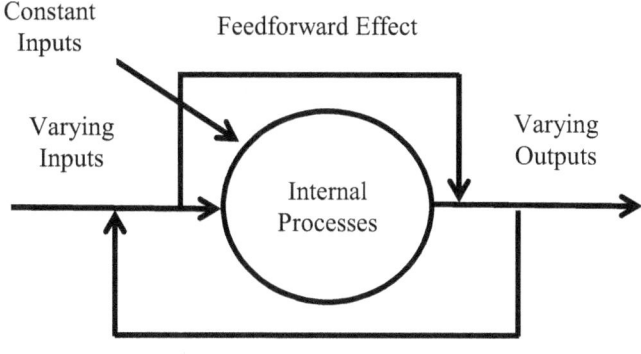

Semi-open Systems

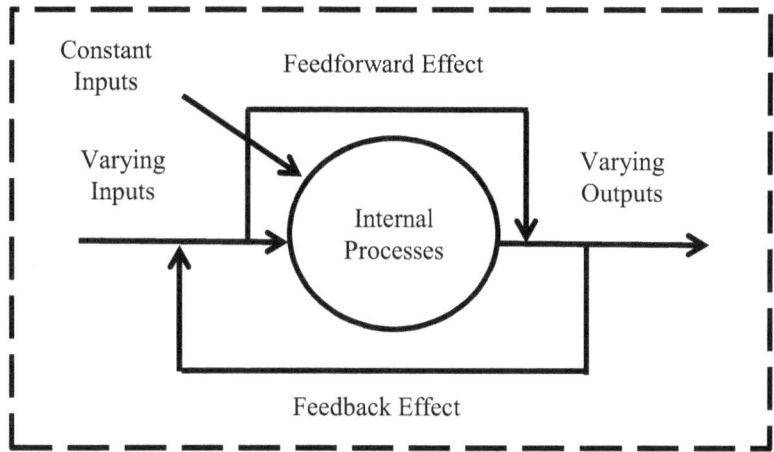

Closed Systems

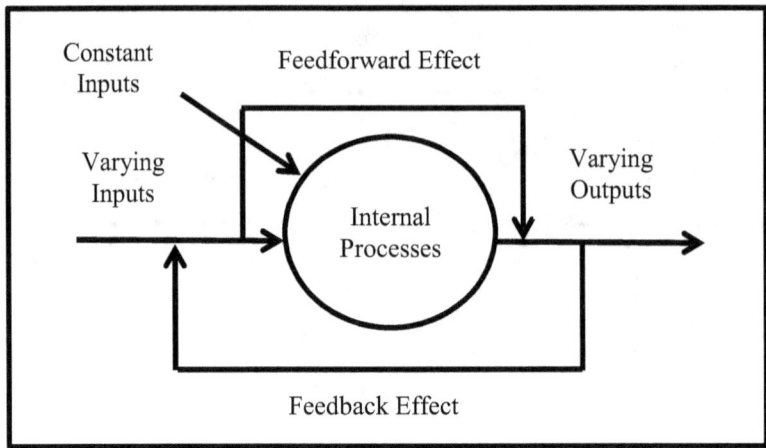

System dynamics and internal processes can be analyzed by specifying one or several inputs or state variables, constant inputs with simple or complex rules governing the state variable/constant interactions, resulting in output variables. Depending on the system type, the general mathematical relationships between input(s) and output(s) variables can be described through discrete time steps (difference equations) or continuous timelines (differential equations)[19].

Mathematical equations governing systems[20] can be linear/nonlinear or difference/differential equations. When inputs or state variables are

---

[19] Also see, **Sayama, Hiroki (2017),** "Introduction to the Modeling and Analysis of Complex Systems," pp. 29. Open SUNY Textbooks, Geneseo, NY.

[20] We should explain some technical definitions for systems.
- **Discrete Systems;** are systems at which the system variables are discrete and not continuous. System's dynamic equations are expressed in terms of discrete or difference equations.
- **Continuous Systems;** are systems at which the system variables are continuous and not discrete. System's dynamic equations are expressed in terms of continuous or differential equations.
- **Linear Systems;** are systems at which the system variables have linear relationships leading to the outputs.
- **Non-linear Systems;** are systems at which system variables have nonlinear relationships leading to outputs.
- **Degree of an Equation;** The highest power of a system's equation..
- **First-order Systems;** are discrete or continuous systems at which system variables are dependent on one past period.

not raised to a power or do not have interacting relationships with other system parameters, we get a linear difference or differential equation. Suppose the inputs or state variables are raised to a power and interact with other system parameters. In that case, we will have a nonlinear difference or differential equation with very complex dynamic output behavior. The input or state variables can also be deterministic, or for a unique given input value, we will get a unique output value. We can also have state variables that are non-deterministic or stochastic. In these situations, our input variables take a probabilistic form with various types of statistical distributions. In such cases, the resulting output variables will also behave in a stochastic and probabilistic manner. Eight possible equations describing different system dynamics are shown in the following table.

Table 1

Linear, Deterministic, Discrete Equations (**LDDI**)	Linear, Deterministic, Differential Equations (**LDDE**)
Nonlinear, Deterministic, Discrete Equations (**NDDI**)	Nonlinear, Deterministic, Differential Equations (**NDDE**)
Linear, Non-Deterministic, Discrete Equations (**LNDI**)	Linear, Non-Deterministic, Differential Equations (**LNDE**)
Nonlinear, Non-Deterministic, Discrete Equations (**NNDI**)	Nonlinear, Non-Deterministic, Differential Equations (**NNDE**)

Simple examples for eight possible types of system dynamic equations are shown below. In these equations, *x* refers to a deterministic or stochastic input or state variable, *t* refers to time, and *a* and *b* are input constants.

✓ **LDDI**
$$x_t = ax_{t-i} \pm b \; for \; i = 1 \ldots . i$$

- **Higher-order Systems;** are discrete or continuous systems at which system variables are dependent on more than one past period.
- **Autonomous Systems;** are systems at which system variables and outputs are not dependent on time or some other external variables.
- **Non-autonomous Systems;** are systems at which system variables and outputs are dependent on time or some other external variables.

In this *ith* order, *1ˢᵗ* degree, Discrete, Autonomous system, we have:

x_t as output and x_{t-i} as input, both deterministic variables

- ✓ **NDDI**
$$x_t = ax_{t-i}(1 \pm x_{t-i}) \pm b \text{ for } i = 1....i$$

In this *ith* order, *2ⁿᵈ* degree, Discrete, Non-linear, Autonomous system, we have:

x_t as output and x_{t-i} as input, both deterministic variables

- ✓ **LNDI**
$$x_t = ax_{t-i} \pm b \text{ for } i = 1....i$$

In this *ith*-order, *1ˢᵗ* degree, Discrete, Autonomous system, we have:

x_t as output and x_{t-i} as input, both stochastic variables

- ✓ **NNDI**
$$x_t = ax_{t-i}(1 \pm x_{t-i}) \pm b \text{ for } i = 1....i$$

In this *ith* order, *2ⁿᵈ* degree, Discrete, Non-linear, Autonomous system, we have:

x_t as output and x_{t-i} as input, both stochastic variables

- ✓ **LDDE**
$$\frac{dx}{dt} = ax \pm b$$

In this *1ˢᵗ* order, *1ˢᵗ* degree, Continuous, Non-autonomous system, we have:

$\frac{dx}{dt}$ as output and x as input, both deterministic variables

- ✓ **NDDE**

$$\frac{dx}{dt} = ax(x \pm 1) \pm b$$

In this *1st* order, *2nd* degree, Continuous, Nonlinear, Non-autonomous system, we have:

$\frac{dx}{dt}$ as output and x as input, both deterministic variables

- ✓ **LNDE**

$$\frac{dx}{dt} = ax \pm b$$

In this *1st* order, *1st* degree, Continuous, Non-autonomous system, we have :

$\frac{dx}{dt}$ as output and x as input, both stochastic variables

- ✓ **NNDE**

$$\frac{dx}{dt} = ax(x \pm 1) \pm b$$

In this *1st* order, *2nd* degree, Continuous, Nonlinear, Non-autonomous system, we have:

$\frac{dx}{dt}$ as output and x as input, both stochastic variables

These equations can be a lot more complicated. In particular, when various variables interact with one another or several equations are coupled together, they become dynamically tricky to understand and, most of the time, mathematically impossible to solve. It is essential to have an excellent conceptual picture of these equations' nature, set them up correctly, and use various mathematical software such as Mathlab, Python, Mathematica, and solve them numerically.

To have the correct mathematical model, one must make sure the following properties of the model are in place:

- *Simplicity;* the model should be simple and understandable. Use Occam's Razor[21] principle.
- *Predictability;* the model should be valid or have good enough predictability power.
- *Robustness;* model's predictions should be reliable, robust, resilient, and insensitive to minor changes to the model's assumptions and other parameters.

Let us look at some basic discrete and continuous systems and their relevant difference or differential equations. What types of systems are described by these dynamic equations assuming that x is deterministic?

✓ $x_t = 2x_{t-1} + 4$

It is a 1st order, 1st degree, Discrete, Linear, Autonomous, deterministic system. x_t is output and x_{t-1} is input.

✓ $x_t = x_{t-1}x_{t-3} + \sin t$

It is a 3rd order, 2nd degree, Discrete, Non-linear, Non-autonomous, deterministic system. x_t is output and x_{t-1}, x_{t-3} and t are inputs.

✓ $x_t = 3x_{t-1} + x_{t-2} t$

It is a 2nd order, 1st degree, Discrete, Non-linear, Non-autonomous, deterministic system. x_t is output and x_{t-1}, x_{t-2} and t are inputs.

✓ $x_t = x_{t-1}x_{t-4} + x_{t-2}t$ where x_{t-i} is stochastic

It is a 4th order, 2nd degree, Discrete, Nonlinear, Non-autonomous, stochastic system. x_t is output and x_{t-1}, x_{t-2}, x_{t-4} and t are inputs.

✓ In mathematical number systems, Fibonacci numbers are a series that are related according to the following equation:

---

[21] **Occam's Razor,** is the principle of choosing the simpler path, solution or model when you have several options.

$$x_t = x_{t-1} + x_{t-2} \text{ where } x_0 = 1, \quad x_1 = 1$$

What type of a number system is described by the Fibonacci series? It is simply a 2nd order, 1st degree, Discrete, Linear, Non-autonomous, deterministic system of numbers. x_t is output and x_{t-1}, x_{t-2} are inputs.

✓ $\frac{dx}{dt} = 2x - 4$

It is a 1st order, 1st degree, Continuous, Linear, Non-autonomous, deterministic system. $\frac{dx}{dt}$ is output, and x is input.

✓ $\frac{dx}{dt} = 3x(x+1) + 4$

It is a 1st order, 2nd degree, Continuous, Nonlinear, Non-autonomous, deterministic system. $\frac{dx}{dt}$ is output, and x is input.

✓ $\frac{d^2x}{dt^2} = 3x^2(x+1) + Sin\ t$

It is a 2nd order, 3rd degree, Continuous, Nonlinear, Non-autonomous, deterministic system. $\frac{d^2x}{dt^2}$ is output, and x and t are inputs.

Let us look at the classical Predator-Prey system and set up the dynamics and mathematical structures involved.

Predator-Prey Systems
In this system that is also known as the Lotka-Volterra system, we assume a species x or prey that has a growth rate equal to:

$$\frac{dx}{dt} = ax \text{ where } a \text{ is the per capita growth rate of } x$$

The prey or x species can grow exponentially as long as there are no predators around. We also have a separate predator species y that would eliminate x. It cannot do so if x is not around and will die over time with the following dying rate:

$$\frac{dy}{dt} = -cy \quad \text{where } c \text{ is the per capita elimination rate of } y$$

We put these two species together and assume an interacting or coupling term of bxy (we could have defined any other interacting term between these two species). For species x, it will have a negative or reducing effect (due to the impact of predation or elimination rate). For species y, it will have a positive or growing effect due to the growth rate effect. The coupled equations describing the dynamics of this closed system are:

$$\frac{dx}{dt} = ax - bxy \quad (growth\ rate\ minus\ elimination\ rate\ for\ x)$$

$$\frac{dy}{dt} = -cy + bxy \quad (elimination\ rate\ plus\ growth\ rate\ for\ y)$$

These two equations describe a coupled, 1st order, 2nd degree, continuous, nonlinear, non-autonomous system. In this system, $\frac{dx}{dt}$ and $\frac{dy}{dt}$ are first-order outputs and x and y are inputs. The solutions for the number of species x and y turns out to be periodic[22] over time. Based on the assumptions for parameters $a, b, c,$ or the coupling and the interacting nature of x and y, we get various versions of this system, analyzed in detail in other studies. Our purpose for mentioning the Predator-Prey model was to understand the nature of more complicated coupled interacting systems.

There are many situations in multi-variable, nonlinear systems (such as complex adaptive systems) where the closed mathematical solutions to the system's equations can be challenging to determine. At other times the value and interplay of the system constant parameters are such that the solutions are many leading to the system dynamics becoming chaotic or undeterminable. These concepts fall under the subject of nonlinear

---

[22] Also see, **Willi-Hans Steeb (2015)**, "The Nonlinear Workbook," 6th Edition, pp. 138. World Scientific Publishing Co. Hackensack, NJ.

dynamics and Causal Inference[23] approaches. The concept of chaotic systems is fascinating and, under certain conditions, can be very pertinent to complex adaptive system analysis[24].

Additionally, we must understand and clarify definitions and differences between Simple, Complicated, Multi-Agent, and Complex Systems before analyzing Complex Adaptive Systems.

Simple Systems (SS)

A simple system[25] is a collection of not more than two or three independent homogeneous elements, components, parts, pieces, nodes, or agents. The elements of a simple system do not influence or have interactions amongst one another. In classical gas dynamics, an Ideal gas assumption of no energy interchange between gas molecules is a simple way to assume that the system is simple and its parts do not interact. We must keep in mind that simple systems have very few elements, and their behavior is linear and non-interactive among elements. Simple systems usually have static equilibrium states (Shayan, 2019).

Complicated Systems (CS)

A complicated system[26] comprises multiple homogeneous bodies of independent elements, components, parts, pieces, nodes, or agents that are a part of a collection. The elements have no or minimal linear interactions amongst one another. If one part or element is removed, the total system's behavior is not affected due to elemental independence, resulting in the system's structure having a high degree of freedom or symmetry. Complex behavior rises when dependencies between elements become relevant. Whenever we have a multi-component homogeneous system with no correlational or contagion effect and interactions among the elements, we will get a non-complex or complicated system. The dynamics of a complicated system can usually be analyzed through its

---

[23] **Causal Inference;** is the reasoning through a cause and effect process that something is likely to be the cause of something else. This mode of reasoning is not based on the results of closed forms of mathematical solutions. It usually is accompanied with statistical or correlational analysis.

[24] Also see, **Steven H. Strogatz (2015)**, "Nonlinear Dynamics and Chaos – With Application to Physics, Biology, Chemistry, and Engineering." Westview Press. Boulder, CO.

[25] **Shayan, S. A. (2019)**,"Understanding Complex Adaptive Systems," Independent Publisher, Amazon.

[26] Ibid.

microstructures or subunits. Due to no or simple linear relations between subunits, one can add up the subunits' behavior through statistical or non-statistical methods and understand the system's total behavior. Complicated systems are more challenging to analyze than simple systems. Statistical methods are practical tools for understanding the dynamics of complicated systems. Complicated systems can have simple and possibly multiple equilibrium states (Shayan, 2019).

Multi-Agent Systems (MAS)

As the interrelationship between multiple elements, parts, nodes, or agents in a complicated system extends and evolves into nonlinear forms, a Multi-Agent System[27] or MAS, with Semi-complex behavior, appears. If the stated features and agent-agent relationships are fully enhanced and grow into having heterogeneity and correlational effects, a Complex System or COM emerges. Complex Systems are a more evolved version of Multi-Agent Systems. The elements have nonlinear interactions among them. If one element is removed, the total system's behavior is affected due to the elemental dependencies, resulting in the system's structure having a lower degree of freedom or symmetry. Multi-Agent or Complex systems can have multiple equilibrium states (Shayan, 2019). As we know it, the universe's galactic structure, with all the interacting stars, galaxies, clusters, and supergalactic clusters in a curved four-dimensional space-time influenced by intense gravitational waves and dark energy, is a perfect example of a very complex system.

Complex Adaptive Systems (CAS)

As the interrelationship between the elements or agents in Multi-Agent Systems (MAS) becomes more heterogeneous, nonlinear with contagion, cooperative, coupling features, network-like interconnected agents with cognitive, learning, corrective abilities can make adaptive and self-organized changes. Holistic emergent structural phenomena appear at this stage, and we have a fully Complex Adaptive System[28] or CAS at work. In such structures and interactive environments, the degree of heterogeneity and internal structural complexity has increased dramatically.

The human brain with a complex network of interacting neurons and individuals' interaction in social media are examples of Complex

---

[27] Ibid.
[28] Ibid.

Adaptive Systems. The progression from a Simple to a highly Complex Adaptive System occurs as the degree of heterogeneity and the relationship between multiple agents increases and develops into nonlinear, interactive, and even cognitive and decision-making modes. Complex Adaptive Systems are based on a structure of highly interrelated network-based agent-agent interactions (Shayan, 2019).

The following table is a brief categorization of different types of systems described so far:

Table 2

	System Variables, Nodes or Agents "N"		
	Discrete Systems		Continuous Systems
	$N \leq 3$	$N \gg 3$	
Linear System	**SS** Linear Oscillator Dynamics RLC Circuit Dynamics	**CS** Molecular Dynamics Statistical Mechanics	**CS** Wave dynamics Viscous Fluid Dynamics
Nonlinear System	**COM** Predator-Prey Dynamics 2 or 3 body Dynamics	**COM or CAS** Swarm or Crowd Dynamics Soccer Game Dynamics AI & Adaptive Neural Network Dynamics	**COM or CAS** Quantum Field Dynamics Financial Market Dynamics Pandemic Dynamics

From nonlinear systems analysis perspectives, determining the stability and continuity of system outputs is essential. As stated before, due to the internal interactive dynamics of highly nonlinear systems, using mathematical equations to find the output stability points can be very difficult. To understand how far or close we are to an unstable, unpredictable, volatile, and chaotic[29] behavior of nonlinear and complex systems' outputs, we can usually rely on two critical parameters. They are; Hurst Exponent and Lyapunov Exponent.

Hurst Exponent

We use this exponent to determine if a network or a system with n nodes or agents could have chaotic, unstable, and unpredictable outputs. The exponent measures the relative variability or unpredictability of a system's output over time. It is done by measuring various system outputs

---

[29] **Chaotic Behavior;** is a long-term unpredictable behavior of a deterministic system with sensitive dependence on initial conditions **(Strogatz, 2015, pp. 331)**.

over time. Next, by choosing different output samples sizes 1 to *i*, we determine each output sample range or R_i (difference between the maximum and minimum value of the output in the sample) and divide them by the standard deviation for each sample or σ_i. We call this range adjusted volatility or RAV for sample size *i*. A more significant sample size *i* leads to a larger R_i and smaller σ_i. According to Hurst, the following power law exists between *RAV* or $\frac{R_i}{\sigma_i}$ and *i*, which is a general form of random walk equation:

$$RAV = \frac{R_i}{\sigma_i} = i^H$$

or

$$\ln RAV = \ln(\frac{R_i}{\sigma_i}) = H \ln i$$

H is the slope of the graph of two logarithms of $\frac{R_i}{\sigma_i}$ and *i* (or sample size). One could do the same analysis for a system with two, three, or m outputs. In higher than one or m dimensional output cases, we will get a Hurst exponent vector with m dimensions. There are three general possibilities for the Hurst exponent's value when dealing with a system's outputs. They are:

- ✓ **If H = 0.5,** the system output behaves as a *"random walk"* or has *"Pure noise."* This situation corresponds to a random walk equation of $RAV = \sqrt{i}$.
- ✓ **If $0 \leq H < 0.5$,** the system output has a tendency to flip flop or fluctuate, or if the system output is increasing, it is more likely to decrease, and if it is decreasing, it is more likely to increase next. We call this output behavior *"anti-persistence random walk"* or a *"Pink noise"* in the system.
- ✓ **If $0.5 < H \leq 1$,** the system output has a tendency to get larger, amplified, more correlated, and volatile in the next step. If the system output is increasing, it is more likely to increase next, and if decreasing, it is more likely to decrease. This behavior tends to lead to unpredictability and catastrophes. We call this output behavior *"persistence random walk"* or a *"Black noise"* in the system. For H above 0.5, we have more correlation or dependencies between system outputs over time. The closer the value of H gets to 1, the more possibility exists to have

output instabilities and observing chaotic system outputs. When observing Hurst exponent system output values close to 1, one can assume nonlinearity and positive feedback, feedforward, and amplifying effects in the system's structure.

Lyapunov Exponent

Assume that we have a time-dependent differentiable variable *x(t)* as input and a function *F(x(t))* as the output for a complex system. We also assume that for every $x(t)$, we will get an output $F(x(t))$. At time zero, we have an initial value $x(t_0)$ and an output value equal to $F(x(t_0))$. If we add a small perturbation and change $x(t_0)$ infinitesimally by $\delta(t_0)$ leading to $x(t_0) + \delta(t_0)$, we will get an output equal to $F(x(t_0) + \delta(t_0))$.

Let us do the same calculations at time *t* for $x(t_t)$ and get $F(x(t_t))$. Again if we change $x(t_t)$ infinitesimally by $\delta(t_t)$ leading to $x(t_t) + \delta(t_t)$, we will get $F(x(t_t) + \delta(t_t))$.

If the infinitesimal change from $\delta(t_0)$ to $\delta(t_t)$ varies through increments of time according to the following power law:

$$\delta(t_t) = \delta(t_0) e^{\lambda_t t_t}$$

We will then define individual Lyapunov exponents as;

$$\lambda_t = \frac{1}{t_t} \ln\left(\frac{\delta(t_t)}{\delta(t_0)}\right)$$

For simplicity, if we have t_t set as equal steps equivalent to *t*, we can then define total Lyapunov exponent as:

$$\lambda \approx \lim_{t \to \infty} \frac{1}{t} \sum_{t=1}^{t=t} \ln\left(\frac{\delta(t_t)}{\delta(t_{t-1})}\right) \quad Total\ for\ t\ equal\ steps$$

Assume $\delta(t_0)$ is our initial change, and it is very small. Lyapunov exponent can be considered the average rate of divergence (positive value) or convergence (negative value) of a system's nearby output results or trajectories. Lyapunov exponent helps quantify a complex system output's closeness to becoming chaotic or volatile, which can be used to detect potential faults, instabilities, unpredictability, and exponential divergence from perturbed initial conditions. *Like the Hurst* exponent, one could calculate the Lyapunov exponent for a system with two, three, or m outputs. *In higher than one or m dimensional output cases, we will get a Lyapunov exponent vector with m dimensions.* There are three general possibilities for the Lyapunov exponent value when dealing with the output of a system. They are as follows:

- ✓ **If $\lambda > 0$,** the average rate of infinitesimal change between two adjacent system outputs is diverging. It means that the exponent λ is positive and δ_t is growing, or the system's output is becoming unstable, unpredictable, and chaotic. This type of output dynamics is usually observed when the system has internal correlational or positive feedback and feedforward structures leading to volatile, *"amplifying,"* unstable, and *"chaotic"* output results over time.
- ✓ **If $\lambda = 0$,** the average rate of infinitesimal change between two adjacent system outputs is constant. It means that the exponent λ is zero and δ_t is constant and equal to δ_0. This type of output dynamics is usually observed when the system is on the *"edge of becoming chaotic."* With a small perturbation, such systems could potentially have a positive λ and become chaotic.
- ✓ **If $\lambda < 0$,** the average rate of infinitesimal change between two adjacent system outputs converges to zero. It means that the exponent λ is negative and δ_t is reducing. This type of system output dynamics is usually observed when internal damping, negative feedback, and feedforward structures lead to stable or attractor (could be a point or periodic) output results over time.

Let us look at a Hurst Exponent problem. Assume that we have the following output observations for R_i and σ_i in various samples i from a complex system. The following table shows all the calculated data and how the exponent is changing. What is the final state of the system?

Calculation of the Hurst Exponent for the output of a complex system

R_i	σ_i	$\dfrac{R_i}{\sigma_i}$	$\ln(\dfrac{R_i}{\sigma_i})$	i	$\ln i$	H
20	5.0	4.0	1.4	10	2.3	0.60
30	4.4	6.8	1.9	20	3.0	0.64
35	4.2	8.3	2.1	30	3.4	0.62
40	4.0	10.0	2.3	40	3.7	0.62
45	3.7	12.2	2.5	50	3.9	0.64
50	3.5	14.3	2.7	60	4.1	0.65
55	3.3	16.7	2.8	70	4.2	0.66
60	3.1	19.4	3.0	80	4.4	0.68
65	2.8	23.2	3.1	90	4.5	0.70
70	2.5	28.0	3.3	100	4.6	0.72

This system's output has a Hurst exponent larger than 0.5 that increases from 0.60 to 0.72 and shows a move from pure noise or random walk towards black noise or persistence, becoming more unstable and chaotic. The final sample of 100 output observations shows a Hurst Exponent equal to 0.72.

For a Lyapunov Exponent problem, we have shown the $\delta(t_i)$ and $\delta(t_{i-1})$ data for a single output observation of a nonlinear system. The following table shows all the necessary data and calculations. What is the total Lyapunov exponent? How do we interpret the behavior of the system output?

Calculation of the Lyapunov Exponent for the output of a complex system

$\delta(t_i)$	$\delta(t_{i-1})$	$\ln(\dfrac{\delta(t_t)}{\delta(t_{t-1})})$	t	$\dfrac{1}{t}$	λ_t
0.105	0.095	0.100	10	0.100	0.010
0.120	0.105	0.134	10	0.100	0.013
0.140	0.120	0.154	10	0.100	0.015
0.165	0.140	0.164	10	0.100	0.016
0.195	0.165	0.167	10	0.100	0.017
0.235	0.195	0.187	10	0.100	0.019
0.290	0.235	0.210	10	0.100	0.021
0.360	0.290	0.216	10	0.100	0.022
0.455	0.360	0.234	10	0.100	0.023
0.575	0.455	0.234	10	0.100	0.023
Sum		1.800			
λ		0.180			

This nonlinear and complex system has a total average Lyapunov exponent larger than zero and equal to 0.180. The Lyapunov exponent in each period is increasing from 0.01 to a maximum of 0.023 in the last period. The system output shows amplification and moving from the edge of chaos towards becoming more unstable and chaotic.

We are now ready to analyze Complex Adaptive Systems or CAS. *It should be clear that CASs can have multiple outputs and inputs, and their internal processes will probably be composed of many nodes or agents. At times these agents could have cognitive abilities (nodes or agents with self-inducing outputs), with highly nonlinear interacting relationships through multiple feedback and feedforward loops leading to many Hamiltonian cycles.*

Questions

Question 1. What type of a system is described by the following dynamic equation assuming that x is deterministic?

$$x_t = x_{t-1} x_{t-2}$$

Answer 1:
It is a 2nd order, 2nd degree, Discrete, Nonlinear, Autonomous, deterministic system. x_t is output and x_{t-1}, x_{t-2} are inputs.

Question 2. What type of a system is described by the following dynamic equation assuming that x is deterministic?

$$x_t = x_{t-1} x_{t-2} x_{t-4} + \sin t^4$$

Answer 2:
It is a 4th order, 4th degree, Discrete, Nonlinear, Non-autonomous, deterministic system. x_t is output and $x_{t-1}, x_{t-2}, x_{t-4}$ and t are inputs.

Question 3. What type of a system is described by the following dynamic equation assuming that x is stochastic?

$$\frac{d^2 x}{dt^2} = 3x^4(x+y)$$

Answer 3:
It is a 2nd order, 5th degree, continuous, Nonlinear, Non-autonomous, stochastic system. $\frac{d^2 x}{dt^2}$ is output, and x and y are inputs.

Question 4. What type of a system is described by the following dynamic equation assuming that x is stochastic?

$$\frac{d^3 x}{dt^3} = x + \cos t$$

Answer 4:
It is a 3rd order, 1st degree, continuous, Nonlinear, Non-autonomous, stochastic system. $\frac{d^3x}{dt^3}$ is output, and **x** and **t** are inputs.

Question 5. What type of a system is described by the following dynamic equations assuming that *x* and *y* are stochastic?

$$\frac{dx}{dt} = ax + by$$

$$\frac{dy}{dt} = -cy + bx$$

Answer 5:
It is a 1st order, 1st degree, coupled, continuous, Linear, Non-autonomous, stochastic system. $\frac{dx}{dt}$ and $\frac{dy}{dt}$ are outputs and *x* and *y* are stochastic inputs.

Question 6. What type of a system is described by the following dynamic Logistic equation assuming that *x* is deterministic?

$$x_{n+1} = rx_n(1 - x_n)$$

Answer 6:
It is a 1st order, 2nd degree, Discrete, Nonlinear, Autonomous, deterministic system. x_{n+1} is the output and x_n is the input. It is an interesting equation for a system since for different values of constant *r;* we will get exciting outcomes for the output x_{n+1}. As r becomes equal or larger than the limiting value of 4.6692… (Feigenbaum constant[30], an

---

[30] **Feigenbaum Constant δ;** is an interesting irrational constant number, which is approximately equal to δ ≈ 4.6692. It is the limiting value of the ratio of the difference between two bifurcation intervals for each period-doubling of a one-dimensional discrete logistic equation. δ is a universal constant number for all functions approaching chaotic behavior through period-doubling, Feigenbaum's number is a fingerprint for detecting instability and chaotic behavior (dynamic behavior that is highly sensitive to initial conditions) in nature. The closer a system's behavior is to the Feigenbaum constant, the more chaotic or sensitive it is to the initial conditions. Also see, **Shayan, S. A. (2020)**,"Fundamental Constants in Mathematics & Physics – Are they universal codes?," Independent Publisher, Amazon.

irrational number), the system output becomes unstable, unpredictable, and chaotic.

Question 7. What type of a system is described by the following dynamic equation assuming that *x* is stochastic?

$$x_{n+1} = 1 - rx_n^2$$

Answer 7:
It is a *1st* order, *2nd* degree, Discrete, Nonlinear, Autonomous, stochastic system. x_{n+1} is the output and x_n is the input.

Question 8. Compare the general features of simple and complex adaptive systems.

Answer 8:
The relative comparison is shown in the following table:

Type of System	Comparing General Features
	General Features
Simple	Small Non-cognitive Parts, Nodes, or Agents. Linear and Simple Equation of the System. Symmetric and Uniform structure. Rigid, Non-flexible fixed equilibrium. Non-unique and Reversible dynamics due to no internal Couplings (no feedback/feedforward loops).
Complex Adaptive	Many Cognitive Parts, Nodes, or Agents. Non-linear and Complex Equation of the System. Asymmetric and Non-Uniform structure. Resilient, Flexible with continuously emerging dynamic equilibrium. Unique and Non-Reversible dynamics due to strong internal Couplings (caused by many feedback/feedforward loops).

Question 9. Provide two examples of nonlinear behavior of two cognitive agents (could be two individuals) interacting.

Answer 9:
When two Chess players or two negotiators are interacting, they have highly nonlinear cognitive interactions. These interactions are very dynamic, with many potential emergent and innovative results.

Question 10. Provide an example of the nonlinear behavior of non-cognitive agents.

Answer 10:
The planets' nonlinear dynamic movement about the sun in the solar system is an example of a non-cognitive, multi-agent, nonlinear behavior.

Question 11. A large international corporation is trying to facilitate and enhance the post-Covid-19 pandemic management and personnel interactions for more safe and effective virtual meetings. Currently, there is no such uniform and organized connectivity with proper controls and monitoring capabilities in place. In line with the post-Covid-19 strategic plan, management has decided to use Zoom on an international scale to achieve such results. What type of a system can this new Zoom-based interaction be considered? Is it Simple, Complicated, MAS, or CAS? Briefly discuss what potential benefits and dangers we see by having such a network and connections being created?

Answer 11:
The planned Zoom-based interaction network program includes multiple changing agent-agent-based connections. Many interacting cognitive agents are connected dynamically on a one-to-one or network structure domestically and internationally. These features will create a system with Complex Adaptive dynamics. The patterns and dynamics created will be CAS-like and must be managed accordingly.

Question 12. Assume that we have the following output observations of R_i and σ_i in various samples i from a CAS. The following table shows all the observations made. Calculate the Hurst exponents? What is the final state of the system?

Calculation of the Hurst Exponent for the output of a complex system

R_i	σ_i	$\dfrac{R_i}{\sigma_i}$	$\ln(\dfrac{R_i}{\sigma_i})$	i	$\ln i$	H
6	2.00			10		
8	1.90			20		
12	1.65			30		
17	1.40			40		
30	1.35			50		
50	1.35			60		
55	1.33			70		
65	1.33			80		
75	1.28			90		
80	1.24			100		

Answer 12:
The following table shows the results.

Calculation of the Hurst Exponent for the output of a complex system

R_i	σ_i	$\dfrac{R_i}{\sigma_i}$	$\ln(\dfrac{R_i}{\sigma_i})$	i	$\ln i$	H
6	2.00	3.00	1.10	10	2.30	0.48
8	1.90	4.21	1.44	20	3.00	0.48
12	1.65	7.27	1.98	30	3.40	0.58
17	1.40	12.14	2.50	40	3.69	0.68
30	1.35	22.22	3.10	50	3.91	0.79
50	1.35	37.04	3.61	60	4.09	0.88
55	1.33	41.35	3.72	70	4.25	0.88
65	1.33	48.87	3.89	80	4.38	0.89
75	1.28	58.59	4.07	90	4.50	0.90
80	1.24	64.52	4.17	100	4.61	0.90

As the sample sizes grow from 10 to 100, we observe a movement from a relative random walk or pure noise behavior of the system outputs with H = 0.48 to close to a chaotic one with H = 0.90 (highly persistence and black noise).

Question 13. Assume that we have the observations of R_i and σ_i for two outputs in various samples i from a complex system. The following table shows all the observations made. Calculate the Hurst exponents for each output? What are the final states for each output? Which output could potentially behave chaotically?

Calculation of the Hurst Exponent for two outputs of a complex system									
R_i^1	σ_i^1	$\ln(\frac{R_i^1}{\sigma_i^1})$	R_i^2	σ_i^2	$\ln(\frac{R_i^2}{\sigma_i^2})$	i	$\ln i$	H_i^1	H_i^2
100	30.00		89	20.00		10			
148	28.00		123	16.00		20			
205	26.00		186	15.00		30			
250	25.00		250	14.00		40			
296	24.00		330	14.00		50			
348	23.00		390	12.00		60			
396	22.00		470	12.00		70			
452	22.00		530	11.00		80			
503	22.00		600	10.00		90			
567	22.00		652	10.00		100			

Answer 13:
The following table shows the results.

Calculation of the Hurst Exponent for two outputs of a complex system									
R_i^1	σ_i^1	$\ln(\frac{R_i^1}{\sigma_i^1})$	R_i^2	σ_i^2	$\ln(\frac{R_i^2}{\sigma_i^2})$	i	$\ln i$	H_i^1	H_i^2
100	30.00	1.20	89	20.00	1.49	10	2.30	0.52	0.65
148	28.00	1.67	123	16.00	2.04	20	3.00	0.56	0.68
205	26.00	2.06	186	15.00	2.52	30	3.40	0.61	0.74
250	25.00	2.30	250	14.00	2.88	40	3.69	0.62	0.78
296	24.00	2.51	330	14.00	3.16	50	3.91	0.64	0.81
348	23.00	2.72	390	12.00	3.48	60	4.09	0.66	0.85
396	22.00	2.89	470	12.00	3.67	70	4.25	0.68	0.86
452	22.00	3.02	530	11.00	3.87	80	4.38	0.69	0.88
503	22.00	3.13	600	10.00	4.09	90	4.50	0.70	0.91
567	22.00	3.25	652	10.00	4.18	100	4.61	0.71	0.91

As the sample sizes grow from 10 to 100, the first output's Hurst coefficient moves from 0.52 (a random walk behavior or pure noise) to 0.71 (persistence or black noise). It is a move towards showing chaotic behavior. Simultaneously, the second output's Hurst exponent moves from 0.65 (smaller persistence and black noise), which already shows a relative sign of chaotic tendencies towards 0.91 (a larger persistence and higher black noise) near chaotic dynamics. The second output could potentially reach chaotic behavior faster. In here the last two Hurst exponents can be shown as a two-dimensional vector as follows:

$$H = \begin{pmatrix} H^1_{last} \\ H^2_{last} \end{pmatrix} = \begin{pmatrix} 0.71 \\ 0.91 \end{pmatrix}$$

This problem could have been extended to a *m* output system with *m* Hurst exponents shown as a *m* dimensional vector.

Question 14. Assume that we have a nonlinear complex system with δ_i and δ_{i-1} observations for a single output. The table shows all the necessary data. Calculate the total Lyapunov exponent? How do we interpret the behavior of the system output?

Calculation of the Lyapunov Exponent for the output of a complex system						
$\delta(t_i)$	$\delta(t_{i-1})$	$\ln(\frac{\delta(t_t)}{\delta(t_{t-1})})$	t	$\frac{1}{t}$	λ_t	
0.095	0.075		5			
0.120	0.095		5			
0.153	0.120		5			
0.180	0.153		5			
0.222	0.180		5			
0.276	0.222		5			
0.335	0.276		5			
0.424	0.335		5			
0.545	0.424		5			
0.693	0.545		5			

Answer 14:
The following table shows the results:

Calculation of the Lyapunov Exponent for the output of a complex system

$\delta(t_i)$	$\delta(t_{i-1})$	$\ln(\frac{\delta(t_t)}{\delta(t_{t-1})})$	t	$\frac{1}{t}$	λ_t
0.095	0.075	0.236	5	0.200	0.047
0.120	0.095	0.234	5	0.200	0.047
0.153	0.120	0.243	5	0.200	0.049
0.180	0.153	0.163	5	0.200	0.033
0.222	0.180	0.210	5	0.200	0.042
0.276	0.222	0.218	5	0.200	0.044
0.335	0.276	0.194	5	0.200	0.039
0.424	0.335	0.236	5	0.200	0.047
0.545	0.424	0.251	5	0.200	0.050
0.693	0.545	0.240	5	0.200	0.048
Sum		2.224			
λ		0.445			

This system has a total Lyapunov exponent larger than zero and equal to 0.445. The Lyapunov exponent, through time, fluctuates between 0.033 and 0.500, showing a general tendency of the system output towards amplification, instability, and chaotic dynamics.

Question 15. Assume that we have a nonlinear complex system with $\delta(t_i)$ and $\delta(t_{i-1})$ observations for two outputs. The following table shows the necessary data. Calculate the two Lyapunov exponents? How do we interpret the behavior of the system outputs?

Calculation of the Lyapunov Exponent for two outputs of a complex system

$\delta(t_i)^1$	$\delta(t_{i-1})^1$	$\ln(\frac{\delta(t_i)^1}{\delta(t_{i-1})^1})$	$\delta(t_i)^2$	$\delta(t_{i-1})^2$	$\ln(\frac{\delta(t_i)^2}{\delta(t_{i-1})^2})$	t	$\frac{1}{t}$	λ_t^1	λ_t^2
0.005	0.004		0.075	0.054		10			
0.007	0.005		0.105	0.075		10			
0.010	0.007		0.165	0.105		10			
0.015	0.010		0.215	0.165		10			
0.023	0.015		0.295	0.215		10			
0.035	0.023		0.385	0.295		10			
0.054	0.035		0.515	0.385		10			
0.085	0.054		0.695	0.515		10			
0.135	0.085		0.981	0.695		10			
0.253	0.135		1.450	0.981		10			
Sum			**Sum**						
λ^1			λ^2						

Answer 15:

The following table shows the results.

Calculation of the Lyapunov Exponent for two outputs of a complex system

$\delta(t_i)^1$	$\delta(t_{i-1})^1$	$\ln(\frac{\delta(t_i)^1}{\delta(t_{i-1})^1})$	$\delta(t_i)^2$	$\delta(t_{i-1})^2$	$\ln(\frac{\delta(t_i)^2}{\delta(t_{i-1})^2})$	t	$\frac{1}{t}$	λ_t^1	λ_t^2
0.005	0.004	0.223	0.075	0.054	0.329	10	0.10	0.022	0.033
0.007	0.005	0.336	0.105	0.075	0.336	10	0.10	0.034	0.034
0.010	0.007	0.357	0.165	0.105	0.452	10	0.10	0.036	0.045
0.015	0.010	0.405	0.215	0.165	0.265	10	0.10	0.041	0.026
0.023	0.015	0.427	0.295	0.215	0.316	10	0.10	0.043	0.032
0.035	0.023	0.420	0.385	0.295	0.266	10	0.10	0.042	0.027
0.054	0.035	0.434	0.515	0.385	0.291	10	0.10	0.043	0.029
0.085	0.054	0.454	0.695	0.515	0.300	10	0.10	0.045	0.030
0.135	0.085	0.463	0.981	0.695	0.345	10	0.10	0.046	0.034
0.253	0.135	0.628	1.450	0.981	0.391	10	0.10	0.063	0.039
Sum		4.147	Sum		3.290				
λ^1		0.415	λ^2		0.329				

The total Lyapunov exponents for the two system outputs are larger than zero and show amplification and chaotic behavior. The first one is larger and is equal to 0.415, and the second one fluctuates and increases through time with an average equal to 0.329. Both exponents show a general tendency towards amplification, instability, and chaotic dynamics. The first output could potentially reach chaotic behavior faster. The two Lyapunov exponents can be shown as a two-dimensional vector as follows:

$$\lambda = \begin{pmatrix} \lambda^1 \\ \lambda^2 \end{pmatrix} = \begin{pmatrix} 0.42 \\ 0.33 \end{pmatrix}$$

This problem could have also been extended to a m output system with m Lyapunov exponents shown as a m dimensional vector.

Complex Adaptive Systems

So far, we have covered the basic concepts of Graphs[31], Networks, and Systems. These concepts could have been analyzed in more detail and at more specialized levels. We covered these concepts at levels that were needed for our purpose to analyze Complex Adaptive Systems. When we combine and apply Network and Systems approaches with the cognitive, multi-agent, nonlinear interactions, we are at the realm of Complex Adaptive Systems or CAS.

From a theoretical perspective, CAS is based on Systems theory built on Network and Graph theories. In other words, Graph theory is the fundamental theory for all CAS.

Graph theory is the study of graphs, which are mathematical structures used to model relations between objects. A graph is made up of vertices (also called nodes, points, or agents) connected by edges (also called links, lines or interactions). A distinction is made between undirected graphs, where edges connect two vertices symmetrically (two-way interaction like a two-way bridge), and directed graphs, where edges link two vertices asymmetrically (one-way interaction like a one-way bridge). Graphs can generally be considered ordered pairs consisting of vertices (nodes, points, or agents) and a set of edges (links, lines, or interactions). The edges are associated with at least two vertices. Graphs are one of the prime objects of study in discrete mathematics. A Graph is generally shown by **G(V, E)**, where V stands for the vertices and E for edges.

It was described previously that network theory deals with studying and analyzing more sophisticated graphs with interrelated vertices and

---

[31] Also see **Steen, Maarten V. (2010),** "Graph Theory and Complex Networks," published by Maarten van Steen.

edges. When we have multi-vertice graphs with multi-edges with symmetric and non-symmetric structures, we are analyzing a network. When a network has multiple inputs and outputs with underlying functions describing how the vertices interact through edges, a systems analytical approach becomes helpful. When we have a system with many cognitive and connected vertices through highly non-linear edges or connections, leading to highly complex feedback and feedforward interactions, we are faced with a Complex Adaptive System or CAS. Modeling CAS is complicated through straightforward mathematical equations. *The most effective approach to analyzing CAS is understanding the holistic dynamics of such systems and using simulation, statistical, and computer modeling.*

Complex Adaptive Systems

Complex Adaptive Systems or CAS are nonlinear systems composed of many interacting vertices or cognitive nodes or agents, without central control, with the ability to adapt to changing environments[32]. In CAS, nodes are replaced with cognitive agents. A network-based structure with multiple cognitive agents is an essential feature of all functioning CAS. Various network structures could affect the complexity of the system's internal reactions, dynamics, and external behavior.

Complex Adaptive Systems are dynamic and evolving multilayer networks or clusters of time-varying agents to agent interaction strengths and types.

Complex Adaptive Systems are characterized by the potential for innovative, emergent new structures, properties, and a high degree of adaptation to external shocks or perturbations. These systems can evolve by random mutation, self-organization transforming their internal agent-agent interactions based on their environments.

We have taken two approaches to analyze CAS. First, we will review, analyze, and quantify the essential holistic features of CAS as best as possible, and some have been described in more detail in the Definition section. Second, we will take a more rigorous and mathematical approach to model CAS.

---

[32] **Shayan, S. A. (2019),**"Understanding Complex Adaptive Systems," Independent Publisher, Amazon.

Holistic Analysis of CAS

To analyze and quantify holistic features of CAS, we mostly rely on Network features covered in the previous chapter in addition to what has been covered in the book entitled Understanding Complex Adaptive Systems (Shayan, 2019). Important predominant holistic features in all Complex Adaptive Systems include;

1. **CAS Type (CT);** we must first determine the type of system that we have. The categories are very similar to the network categories defined before. From a network point of view, complex adaptive systems are dynamic and evolving multilayer networks with time-varying different agent-agent interaction strengths (including feedback and feedforward links) and types. Different categories of CAS include;
    a. **Physical CAS (PC);** have physical components or agents (i.e., water, gas, oil pipelines, and electric grid distributions) with physical links between agents or control centers, such as roads, pipelines, or fiber-optic networks. These are systems with real physical agents or components and links that have specific functions.
    b. **Logical CAS (LC);** have logical or thinking agents (i.e., large connected parallel computer programs) with information flow as possible links.
    c. **Information CAS (IC);** have information points as their agents (information flow through intranets, internet, and wireless telephones), and the links are through wifi, internet, fiber optics, radars, and satellites.
    d. **Blended CAS (BC);** any blend or combination of the other CASs. Most CASs fall into this category.
2. **The number of Agents (NOA);** many agents interacting can significantly impact a system's complexity.
3. **Degree of Non-homogeneity of Agents (DNHA);** the more homogeneous the agents are, the easier it is to understand and model a system. Non-homogeneity, diversity, and variations in agent types in a system add to the output complexity and dynamics. One way to measure this feature is to determine the number of non-homogeneous agents or their clusters in the system. A higher number of non-homogeneous agent clusters can lead to higher complex output dynamics.

4. **Average Number of Connections per Agents (ANCA);** this is the average number of links agents have with each other in the whole system. This general feature will affect the intensity of interactions and the complexity of agent-to-agent interrelationships. A higher value for this holistic attribute can increase systems complexity and structure and help the system become more stable and face more alternatives for system dynamic stability, resilience, and flexibility.
5. **Average Number of In Connections per Agent (ANICA);** this is the average number of links with an agent as their targets. The calculation is done for each agent and on an average basis for the whole system. The higher the ANICA value is, the more *agents act as a sink*.
6. **Average Number of Out Connections per Agent (ANOCA);** this is the average number of links with an agent as their source. The calculation is done for each agent and on an average basis for the whole system. The higher the ANOCA value is, the more *agents act as a source*.
7. **Agent Coupling Strengths (ACS);** over and above the number of agent to agent links in a system and the existing agent clusters, the *weight or coupling* strengths for each agent to agent link are essential. Not all links carry the same impact, weight, or coupling strength with other agents. It is similar to the concept of the interaction coupling strength matrix parameter $I_{ij}(t)$ in networks. In CAS, we define interaction coupling strength C_{ij}^λ for interaction type λ (i.e., gravitational, electromagnetic, flow, correlational), that can vary through time between agents i and j. This parameter can be specified for each sub-clusters 1 to ε and can vary through time as follows:

$C_{ij}^{\lambda\varepsilon}(t)$ for cluster ε from 1 to ε and interaction type λ

In general, for a system with n agents and agent-agent interaction type λ, the interaction coupling strength matrix parameter $C_{ij}^\lambda(t)$ can be scalar (a number), vector (a number with direction), or a function (such as gravitational, electromagnetic, correlational, to name a few). It will have a matrix of n*n interaction coupling strength described as follows:

$$C^\lambda(t) = \begin{pmatrix} C_{11}^\lambda(t) & \cdots & C_{1n}^\lambda(t) \\ \vdots & \ddots & \vdots \\ C_{n1}^\lambda(t) & \cdots & C_{nn}^\lambda(t) \end{pmatrix}$$

This matrix can dynamically change through time, and the change can mathematically be represented as:

$$\frac{dC^\lambda(t)}{dt} = \begin{pmatrix} \frac{dC_{11}^\lambda}{dt} & \cdots & \frac{dC_{1n}^\lambda}{dt} \\ \vdots & \ddots & \vdots \\ \frac{dC_{n1}^\lambda}{dt} & \cdots & \frac{dC_{nn}^\lambda}{dt} \end{pmatrix}$$

The matrix can also be written for each sub-cluster ε separately. The equation will then be as follows:

$$C^{\lambda\varepsilon}(t) = \begin{pmatrix} C_{11}^{\lambda\varepsilon}(t) & \cdots & C_{1n}^{\lambda\varepsilon}(t) \\ \vdots & \ddots & \vdots \\ C_{n1}^{\lambda\varepsilon}(t) & \cdots & C_{nn}^{\lambda\varepsilon}(t) \end{pmatrix}$$

The sub-cluster matrix can also dynamically change through time and can mathematically be represented as:

$$\frac{dC^{\lambda\varepsilon}(t)}{dt} = \begin{pmatrix} \frac{dC_{11}^{\lambda\varepsilon}}{dt} & \cdots & \frac{dC_{1n}^{\lambda\varepsilon}}{dt} \\ \vdots & \ddots & \vdots \\ \frac{dC_{n1}^{\lambda\varepsilon}}{dt} & \cdots & \frac{dC_{nn}^{\lambda\varepsilon}}{dt} \end{pmatrix}$$

By analyzing the interaction coupling strength matrix $C^{\lambda\varepsilon}(t)$, the internal dynamics of a CAS and its components can be understood.

8. **The number of Hamiltonian Cycles[33] (NHC);** Similar to Network analysis, if we start from one agent and go through possible connections and can come back to the same agent

---

[33] **Hamiltonian Cycle;** is a mathematical concept in Graph Theory.

(which usually happens when there are feedback and feedforward connections) without having to go through an agent twice, we have a Hamiltonian cycle in our CAS structure. We can have many Hamiltonian cycles in a given CAS or its underlying clusters. The more multiple and two-way feedback and feedforward connections there are among agents, the more Hamiltonian cycles can be detected. More Hamiltonian cycles add to the complexity of the system and its dynamic behavior. We can use the number of Hamiltonian cycles to understand how complex the system can potentially behave and where the sources of complexity are hidden in the structure. *One can even think about adding or eliminating agent-to-agent links to create or delete Hamiltonian cycles to increase or decrease the system's complexity.* A straightforward way to represent a Hamiltonian cycle is to show the series of agents in progression in each cycle, such as; [agent1, agent3, agent5, agent7, agent1]. The first and last agents in every Hamiltonian cycle should be the same, and the coupling effects must be for links with an agent as their source.

9. **Average System Information Entropy (ASIP);** Equivalent to Network analysis, if each agent in the system behaves stochastically and shows independent random behavior, then its average randomness and behavior uncertainty can be calculated by Shannon's information Entropy measure. If we have a system with n agents, where agent *i* has state *j* with the probability of observing agent *i* in state *j* equal to p_j^i (the cumulative probability of agent *i* in all *j* states will have to be equal to 1), then for each agent, we can calculate information entropy[34] S_i equal to:

$$S_i = -\sum_{j=1}^{j=j} p_j^i \ln p_j^i$$

---

[34] **Shannon's Entropy;** is a concept used in information theory to measure and calculate the variability and uncertainty inherent in the information, outcome or behavior of a system. It was initially presented by Claude Shannon in 1948 as Information Entropy.

And for the total system, an average can be defined as;

$$ASIP = -(\sum_{i=1}^{i=n} \sum_{j=1}^{j=j} p_j^i \ln p_j^i)/n$$

For each agent in the system, the summation of probabilities of observing each state or p_j^i must add up to 1 or:

$$\sum_{j=1}^{j=j} p_j^i = 1$$

As the individual probability for an agent i or p_j^i becomes more significant and closer to 1, S_i gets closer to 0 (meaning that less randomness or more structure exists) or less information entropy per agent. This reasoning is valid if all agents' probabilities become more significant and closer to 1, or *ASIP* gets closer to 0, resulting in more dynamic certainty and less average information entropy for the system. The opposite reasoning is also valid. If p_j^i gets smaller and close to 0, or S_i gets larger, we get less dynamic certainty (more randomness or less structure) or more average information entropy. *Average System Information Entropy or ASIP indicates how much variation, randomness, and uncertainty we should expect from a system's dynamic behavior. The higher the ASIP is, the more randomness and uncertainty the behavior of a system will present. Lower ASIP indicates more structure, memory, cognition, specialization, and certainty in the system's behavior.*

10. **Average System Information Energy (ASIG);**
    CASs are nonlinear, dynamic, and cognitive systems that interact with their environments through information absorption, learning, memorizing, readjusting, restructuring and re-equilibrating. More learning, maturing, and additional negative feedback and feedforward loops are created through these adjustments and changes. The negative feedback and feedforward loops have stabilizing, dampening, calming, anti-volatility effects and are usually the main contributors to increased intelligence and cognition. The structural changes

and deepening of the interactions create additional information contents generating additional complexities incorporating new memory, intelligence, cognitive, specialization clusters, and learning abilities. Through this evolutionary and developmental process, the complexities of the system structure increase. More specialized clusters are then created, which causes the information content or energy encapsulated in the structure to increase. We have defined a parameter called Average System Information Energy or ASIG to capture this critical feature of CAS. If we have a system with n agents interacting nonlinearly, with negative feedback and feedforward loops, where agent *i* has total information energy of e_i, then for all n agents, we can calculate total system information energy equal to E_T as follows:

$$E_T = \sum_{i=1}^{i=n} e_i$$

We will breakdown each agent's total information energy into the following two parts:

- *Information Energy that is actualized through the links and interactions of each agent with other agents.* This actualized information energy is highly dependent on CAS parameters such as ANCA, ACS, and NHC. We will abbreviate this energy as e_{ia}.
- *Information Energy that is not actualized through existing links and interactions and exists as the information energy potential in each agent.* This type of potential energy is highly dependent on the *cognitive* levels of each agent. It can lead to new links, interactions, adaptations, collaborations, coordination, cooperation, coupling, innovation, emergent and synergistic behavior among agents. We will abbreviate this energy as e_{ip}.

We can therefore show that our total system information energy is equal to:

$$E_T = \sum_{i=1}^{i=n}(e_{ia} + e_{ip}) \quad for\ n\ agents$$

Suppose we have a CAS with many agents (such as swarms), with smooth and continuous system transitional behavior and changes (this is a limiting assumption). In that case, we can then define a total information energy Lagrange Equation[35] for this system as follows:

$$L_T = \sum_{i=1}^{i=n}(e_{ia} - e_{ip}) \quad for\ n\ agents$$

For the more mathematically sophisticated readers, the utilization of multi-dimensional Euler-Lagrange concepts and the Principle of Least Action[36] in CAS can lead to many interesting analytical and mathematical conclusions, which for now is beyond the scope of this book. We will leave this for interested readers to pursue. For the total average information energy of the system, we get:

$$ASIG = \sum_{i=1}^{i=n}(e_{ia} + e_{ip})/n$$

And if we define;

$$Aea = \sum_{i=1}^{i=n}(e_{ia})/n$$

---

[35] **Lagrange Equation,** we refer the readers to the concepts of Euler-Lagrange equation, optimization principles using variational calculus, Principle of Least Action and Lagrangian mechanics. They all deal with the general methodology of finding the optimum or stationary states of a multi-dimensional many body system.

[36] **Principle of Least Action,** also known as the Principle of Stationary Action is an important and axiomatic principle explaining the dynamic behavior of all systems in nature. It states that for any system that moves from one state to another, out of many possible paths or trajectories, there exists only one optimal path which the system would take. The application of this principle in physics lead to Newtonian, Lagrangian, Hamiltonian, Quantum Mechanical and even the General Relativity equations of motion. Aslo see: https://en.wikipedia.org/wiki/Stationary_Action_Principle.

$$Aep = \sum_{i=1}^{i=n} (e_{ip}) / n$$

Then,

$$ASIG = Aea + Aep$$

From the above definitions, "Aea" can be defined as the average actualized/interactive information energy and "Aep" as the average potential/cognitive information energy of a CAS. ASIG is strongly dependent on NOA, ANCA, ACS, and NHC. Average System Information Energy gives us an indication of the degree of potential cognitive abilities and internal structural complexity that exists through nonlinearities, negative feedback, and feedforward loops (which could also be detected through the number of Hamiltonian Cycles). The structure's negative loops allow the system to have more structural deepening, sophistication, and information content (think of it as increased maturity, intelligence, and training of the system) and become more specialized, resilient, and adaptable. Higher ASIG allows the system to adapt innovatively, adjust, learn, train, memorize, specialize (through formed clusters), be flexible, re-stabilize, and have emergent behavior when faced with perturbations, shocks, and new information. *As ASIG goes up, more structure, memory, intelligence, and certainty are created in the system, and with this added training and encapsulated energy, ASIP goes down. ASIG and ASIP are linked and behave oppositely. Higher ASIG leads to lower ASIP, and lower ASIG leads to higher ASIP.* We can also show ASIG in terms of the following function:

$$ASIG = F(NOA, ANCA, ACS, NHC, \frac{1}{ASIP})$$

Increasing ASIG or decreasing ASIP of a system (meaning injecting information energy into a system) changes the structural randomness, homogeneity, and symmetry[37] towards structural deepening, order, asymmetry, heterogeneity, sophistication, and increase in emergent behavior with positive synergistic[38] effects.

11. **System Dynamics (SD);** understanding the flow through the agent to agent links and determining the effects of coupling strengths between them, all the way through the cluster to cluster interactions and the whole system is critical. In CAS, the state of each agent and the interaction strength matrix can dynamically change through time. It can lead to co-interacting and co-evolving[39] phenomena between agents and their respective interactions, which can be expressed as a single or sets of coupled nonlinear difference or partial differential equations (nonlinear because we deal with CAS). Sometimes, due to the inability to determine a system's equations, statistical relationships describe the input/output dynamics. It is essential to understand the nature of the equations that describes the dynamics of a system. To analyze and gauge the chaotic nature, we could also utilize the Hurst and Lyapunov exponents as described before.

For now and for analytical purposes, we will use the following CAS features:

---

[37] **Symmetry;** through a complexity lens, it is related to the structural indifference of a system when exposed to some form of transformation. If the system's physical structure and mathematical or informational outputs stay the same after going through a particular transformation, then it is symmetric with regards to that transformation.

[38] **Synergy;** in Greek it means working together. Synergy is a phenomenon or process where the elements, components, parts, pieces, or cognitive or non-cognitive agents in a system will interact and cooperate advantageously for a final outcome or cause that has a combined effect greater than the sum of the separate effects. When extra value or result is created through the dynamic interaction of the components, we observe synergies. This happens when the whole is greater than the sum of its parts. It is a property of certain nonlinear systems with unique synergy creating features. Synergy is positive (Positive Synergy) when the combined effect is greater than the sum of the separate effects and negative (Negative Synergy) when combined effect is less than the sum of the separate effects. Positive Synergy is desirable, and Negative Synergy is not. The Average System Information Energy or ASIG gets larger with positive synergy and smaller with negative synergy.

[39] **Co-evolving;** this phenomena among agents happens when agents have mutual understanding and adaptation amongst one another.

1. CAS Type (CT)
2. Number of Agents (NOA)
3. Non-homogeneity of Agents (NHA)
4. Average Number of Connections per Agents (ANCA)
5. Average Number of In Connections per Agent (ANICA)
6. Average Number of Out Connections per Agent (ANOCA)
7. Agent Coupling Strengths (ACS)
8. Number of Hamiltonian Cycles (NHC)
9. Average System Information Entropy (ASIP)
10. Average System Information Energy (ASIG);
11. System Dynamics (SD);

Let us look at some examples.

Example 1:

How would ASIG, ASIP, and structural symmetries be affected in the following functions?
- Computer programming
- Musical DVD recording
- Saving data on a hard drive
- Mathematical Equations
- Movie production
- Painting
- Building a plant
- Building a robot
- Writing a book

Answer 1

Function	ASIG	ASIP	Symmetry
Computer programming	Increases	Decreases	Decreases
Musical DVD recording	Increases	Decreases	Decreases
Formatting a hard drive	Decrease	Increase	Increase
Writing mathematical equations	Increases	Decreases	Decreases
Movie production	Increases	Decreases	Decreases
Painting	Increases	Decreases	Decreases
Demolishing a building	Decrease	Increase	Increase
Building a robot	Increases	Decreases	Decreases
Writing a book	Increases	Decreases	Decreases

Example 2:

What type of CAS would the following functions be categorized?

Function	CAS Type
Computer programming	
Musical DVD recording	
Formatting a hard drive	
Movie production	
Painting	
Demolishing a building	
Building a robot	
Writing a book	

Answer 2

Function	CAS Type
Computer programming	Blended (Logical, Information, physical)
Musical DVD recording	Blended (Information, physical)
Formatting a hard drive	Blended (Information, physical)
Movie production	Blended (Information, physical)
Painting	Physical
Demolishing a building	Physical
Building a robot	Blended (Logical, Information, physical)
Writing a book	Blended (Information, physical)

Example 3:

How would Hamiltonian cycles show up in the interaction coupling strength matrix of a complex adaptive system?

Answer 3

When we have Hamiltonian cycles in our CAS, we will have feedback and feedforward loops. These loops lead to multiple out and in connections per system agent. If we look at the following CAS with and without feedback loops, the two Hamiltonian cycles cause the interaction coupling strength matrix to change from triangular to asymmetric.

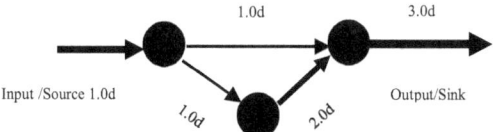

$$C(t) = \begin{pmatrix} 0 & 1d & 1d \\ 0 & 0 & 2d \\ 0 & 0 & 3d \end{pmatrix} \text{ this is a traingular matrix}$$

And with a feedback loop,

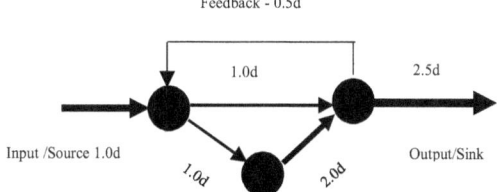

$$C(t) = \begin{pmatrix} 0 & 1d & 1d \\ 0 & 0 & 2d \\ -0.5d & 0 & 3d \end{pmatrix} \text{ this is an asymmetric matix}$$

Example 4:

How would a cognitive agent in a CAS affect itself in the system's interaction coupling strength matrix?

Answer 4

When we do not have agents affecting themselves or having feedback, the interaction coupling strength matrix diagonal will be zero. If agents are cognitive, have learning abilities, or provide feedback to themselves, the matrix's diagonal will have values. These values will depend on the self-feedback strengths of each agent.

Example 5:

Assume that we have a manufacturing production flow system with the internal structure shown below. Let us also assume that in this semi-open system, all agent-agent (production-production) interactions are different and create a final output equal to a function related to a linear combination effect (for now, this is not considered a CAS) of the agent-agent interactions. Agents and links or flows (shown as A, 2A, or 3A) are fixed and deterministic. Determine the system's internal network classification, attributes, and interaction coupling strength matrix.

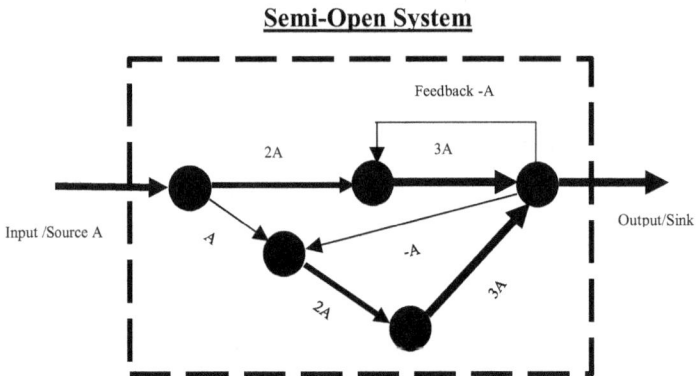

Semi-Open System

Answer 5

For the system internal agent-agent (node-node) network classification, attributes and the interaction coupling strength matrix we have:

- **System Graph:**   Shown above
- **CAS Type:**   Physical
- **System Attributes:**
  - ✓ NOA   5 agents
  - ✓ DNHA   Non-homogeneous
  - ✓ ANCA   9 links with an avg. 1.8 per agent including the source and sink
  - ✓ ANICA   8 links with an avg. 1.6 per agent
  - ✓ ANOCA   8 links with an avg. 1.6 per agent
  - ✓ ACS   Shown on the graph (A, 2A, 3A)
  - ✓ NHC   5

- ✓ ASIP   0 with no variations in node behaviors
  - ✓ SD   Shown on the structure

- **Input/Source:**   A
- **Output/Sink Result:** Agent5 output $3A + 3A - A - A = 4A$
- **Trigger Parameter Vector:**

$$T(t) = \begin{pmatrix} T_1 = 2A + A = 3A \\ T_2 = 2A \\ T_3 = 2A \\ T_4 = 3A \\ T_5 = 3A + 3A - A - A = 4A \end{pmatrix}$$

$$\frac{dT(t)}{dt} = \begin{pmatrix} 0 \\ 0 \\ 0 \\ 0 \\ 0 \end{pmatrix}$$

- **Interaction Coupling Strength Matrix:**

$$C(t) = \begin{pmatrix} 0 & 1A & 2A & 0 & 0 \\ 0 & 0 & 0 & 2A & 0 \\ 0 & 0 & 0 & 0 & 3A \\ 0 & 0 & 0 & 0 & 3A \\ 0 & -A & -A & 0 & 6A \end{pmatrix}$$

$$\frac{dC(t)}{dt} = \begin{pmatrix} 0 & 0 & 0 & 0 & 0 \\ 0 & 0 & 0 & 0 & 0 \\ 0 & 0 & 0 & 0 & 0 \\ 0 & 0 & 0 & 0 & 0 \\ 0 & 0 & 0 & 0 & 0 \end{pmatrix}$$

Example 6:

Assume that we have the same manufacturing production flow system with nonlinearity in the previous example's internal structure, as shown below. Let us assume that all agent-agent (production-production) interactions are different in this semi-open system and create a final output equal to a function related to a linear cumulative effect (this is considered a CAS) of the agent-agent nonlinear interactions. Agents and links or flows are all fixed and deterministic. Determine the system's internal network classification, attributes, and interaction coupling strength matrix.

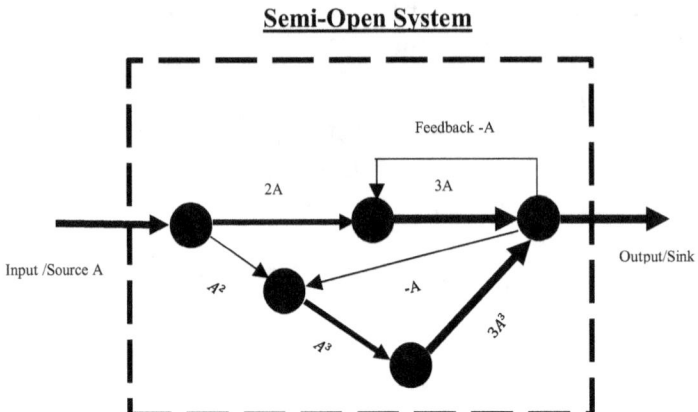

Answer 6

For the system internal agent-agent (node-node) network classification, attributes and the interaction coupling strength matrix we have:

- **System Graph:** Shown above
- **CAS Type:** Physical
- **System Attributes:**
  - ✓ NOA — 5 agents
  - ✓ DNHA — Non-homogeneous
  - ✓ ANCA — 9 links with an avg. 1.8 per agent including the source and sink
  - ✓ ANICA — 8 links with an avg. 1.6 per agent
  - ✓ ANOCA — 8 links with an avg. 1.6 per agent
  - ✓ ACS — Shown on the graph (A, 2A, 3A, A^2, A^3, 3A^2)
  - ✓ NHC — 5
  - ✓ ASIP — 0 with no variations in node behaviors
  - ✓ SD — Shown on the structure

- **Input/Source:** A
- **Output/Sink Result:** Agent5 output $3A^3 + 3A - A - A = A(3A^2 + 1)$
- **Trigger Parameter Vector:**

$$T(t) = \begin{pmatrix} T_1 = 2A + A^2 = A(A+2) \\ T_2 = A^3 \\ T_3 = 3A \\ T_4 = 3A^3 \\ T_5 = 3A^3 + 3A - A - A = A(3A^2 + 1) \end{pmatrix}$$

$$\frac{dT(t)}{dt} = \begin{pmatrix} 0 \\ 0 \\ 0 \\ 0 \\ 0 \\ 0 \end{pmatrix}$$

- **Interaction Coupling Strength Matrix:**

$$C(t) = \begin{pmatrix} 0 & A^2 & 2A & 0 & 0 \\ 0 & 0 & 0 & A^3 & 0 \\ 0 & 0 & 0 & 0 & 3A \\ 0 & 0 & 0 & 0 & 3A^3 \\ 0 & -A & -A & 0 & (3A^3 + 3A) \end{pmatrix}$$

$$\frac{dC(t)}{dt} = \begin{pmatrix} 0 & 0 & 0 & 0 & 0 \\ 0 & 0 & 0 & 0 & 0 \\ 0 & 0 & 0 & 0 & 0 \\ 0 & 0 & 0 & 0 & 0 \\ 0 & 0 & 0 & 0 & 0 \end{pmatrix}$$

Example 7:

If in the previous question the probability distribution for the output source for each agent were independent and varied through time for each state a (this system can be considered as CAS) as follows:

$$A(a) = 1 + e^{-at} \quad \text{and} \quad \sum_{a=1}^{a=a} 1 + e^{-at} = 1$$

Find out the agent Trigger Parameters, Interaction Parameter Coupling Strength Matrix, their changes through time, and the Average System Information Entropy or ASIP.

Answer 7

We get,

- **Trigger Parameter Vector:**

$$T(t) = \begin{pmatrix} T_1 = (1+e^{-at})(3+e^{-at}) \\ T_2 = (1+e^{-at})^3 \\ T_3 = 3(1+e^{-at}) \\ T_4 = 3(1+e^{-at})^3 \\ T_5 = 3(1+e^{-at})^3 + (1+e^{-at}) \end{pmatrix}$$

$$\frac{dT(t)}{dt} = \begin{pmatrix} \frac{dT_1}{dt} \\ \frac{dT_2}{dt} \\ \frac{dT_3}{dt} \\ \frac{dT_4}{dt} \\ \frac{dT_5}{dt} \end{pmatrix}$$

Note: We have not taken the derivatives. We do not intend to practice calculus. Interested readers can do this as an exercise.

- **Interaction Coupling Strength Matrix:**

$$C(t) = \begin{pmatrix} 0 & (1+e^{-at})^2 & 2(1+e^{-at}) & 0 & 0 \\ 0 & 0 & 0 & (1+e^{-at})^3 & 0 \\ 0 & 0 & 0 & 0 & 3(1+e^{-at}) \\ 0 & 0 & 0 & 0 & 3(1+e^{-at})^3 \\ 0 & -1(1+e^{-at}) & -1(1+e^{-at}) & 0 & 3(1+e^{-at})^3 + 3(1+e^{-at}) \end{pmatrix}$$

$$\frac{dC(t)}{dt} = \begin{pmatrix} 0 & -2ae^{-at}(1+e^{-at}) & -2ae^{-at} & 0 & 0 \\ 0 & 0 & 0 & -3ae^{-at}(1+e^{-at})^2 & 0 \\ 0 & 0 & 0 & 0 & -3ae^{-at} \\ 0 & 0 & 0 & 0 & -9ae^{-at}(1+e^{-at})^2 \\ 0 & ae^{-at} & ae^{-at} & 0 & -9ae^{-at}(1+e^{-at})^2 - 3ae^{-at} \end{pmatrix}$$

- ASIP equals to:

$$S_1 = -\sum_{a=1}^{a=a}(1+e^{-at})\ln(1+e^{-at})$$

$$S_2 = -\sum_{a=1}^{a=a}(1+e^{-at})\ln(1+e^{-at})$$

$$S_3 = -\sum_{a=1}^{a=a}(1+e^{-at})\ln(1+e^{-at})$$

$$S_4 = -\sum_{a=1}^{a=a}(1+e^{-at})\ln(1+e^{-at})$$

$$S_5 = -\sum_{a=1}^{a=a}(1+e^{-at})\ln(1+e^{-at})$$

$$ASIP = \frac{S_1+S_2+S_3+S_4+S_5}{5} = -\sum_{a=1}^{a=a}(1+e^{-at})\ln(1+e^{-at})$$

Example 8:

In the previous question, find out the agent Trigger Parameters, Interaction Coupling Strength Matrix, their changes through time, and the Average System Information Entropy or ASIP as t becomes very large or goes to infinity (this system can be considered as CAS).

Answer 8

We get,

$$\lim_{t\to\infty} T(t) = \begin{pmatrix} T_1 = 3 \\ T_2 = 1 \\ T_3 = 3 \\ T_4 = 3 \\ T_5 = 4 \end{pmatrix}$$

$$\lim_{t \to \infty} \frac{d\boldsymbol{T}(t)}{dt} = \begin{pmatrix} 0 \\ 0 \\ 0 \\ 0 \\ 0 \end{pmatrix}$$

- **Interaction Coupling Strength Matrix:**

$$\lim_{t \to \infty} \boldsymbol{C}(t) = \begin{pmatrix} 0 & 1 & 2 & 0 & 0 \\ 0 & 0 & 0 & 1 & 0 \\ 0 & 0 & 0 & 0 & 3 \\ 0 & 0 & 0 & 0 & 3 \\ 0 & -1 & -1 & 0 & 6 \end{pmatrix}$$

$$\lim_{t \to \infty} \frac{d\boldsymbol{C}(t)}{dt} = \begin{pmatrix} 0 & 0 & 0 & 0 & 0 \\ 0 & 0 & 0 & 0 & 0 \\ 0 & 0 & 0 & 0 & 0 \\ 0 & 0 & 0 & 0 & 0 \\ 0 & 0 & 0 & 0 & 0 \end{pmatrix}$$

- **ASIP equals to:**

$$\lim_{t \to \infty} S_1 = 0$$

$$\lim_{t \to \infty} S_2 = 0$$

$$\lim_{t \to \infty} S_3 = 0$$

$$\lim_{t \to \infty} S_4 = 0$$

$$\lim_{t \to \infty} S_5 = 0$$

$$\lim_{t \to \infty} ASIP = \lim_{t \to \infty} \frac{S_1 + S_2 + S_3 + S_4 + S_5}{5} = 0$$

Example 9:

In the previous question, find out the agent Trigger Parameters, Interaction Coupling Strength Matrix, their changes through time, and the Average System Information Entropy or ASIP as t goes to zero and compare this state with t becoming very large or going to infinity (this system can be considered as CAS). Assuming no change is made to the CAS agents or their links, determine which state has higher and lower average system information energy or ASIG.

Answer 9

We get,

$$\lim_{t \to 0} \mathbf{T}(t) = \begin{pmatrix} T_1 = 2*4 = 8 \\ T_2 = 2*2*2 = 8 \\ T_3 = 3*2 = 6 \\ T_4 = 3*8 = 24 \\ T_5 = 24+2 = 26 \end{pmatrix}$$

$$\lim_{t \to 0} \frac{d\mathbf{T}(t)}{dt} = \lim_{t \to 0} \begin{pmatrix} 0 \\ 0 \\ 0 \\ 0 \\ 0 \end{pmatrix}$$

- **Interaction Coupling Strength Matrix:**

$$\lim_{t \to 0} \mathbf{C}(t) = \begin{pmatrix} 0 & 4 & 4 & 0 & 0 \\ 0 & 0 & 0 & 8 & 0 \\ 0 & 0 & 0 & 0 & 6 \\ 0 & 0 & 0 & 0 & 24 \\ 0 & -2 & -2 & 0 & 30 \end{pmatrix}$$

$$\lim_{t \to 0} \frac{d\mathbf{C}(t)}{dt} = \begin{pmatrix} 0 & -4a & -2a & 0 & 0 \\ 0 & 0 & 0 & -12a & 0 \\ 0 & 0 & 0 & 0 & -3a \\ 0 & 0 & 0 & 0 & -36a \\ 0 & a & a & 0 & -39a \end{pmatrix}$$

- **ASIP equals to:**

$$\lim_{t \to 0} S_1 = -2\ln 2$$

$$\lim_{t \to 0} S_2 = -2\ln 2$$

$$\lim_{t \to 0} S_3 = -2\ln 2$$

$$\lim_{t \to 0} S_4 = -2\ln 2$$

$$\lim_{t \to 0} S_5 = -2 \ln 2$$

$$\lim_{t \to 0} ASIP = \lim_{t \to 0} \frac{S_1 + S_2 + S_3 + S_4 + S_5}{5} = -2 \ln 2$$

To compare the situation for both ASIP and ASIG when t equals zero and infinity, we look at the following table:

CAS parameter	$t \to 0$	$t \to \infty$
ASIP	smaller	larger
ASIG	larger	smaller

As time goes on (or goes towards infinity), the average system information entropy (or information uncertainty) and structural symmetries become larger. It means that the structure becomes more homogeneous while the average system information energy is reduced. As we lose information structure, we lose information energy.

It will be challenging for systems with many agents to determine the agent-agent interactions' exact nature and the possible links. We can analyze the dynamics in such systems by looking at the relationship between the output and input data using neural networks, statistical, econometric, or any other simulation and correlational analysis[40].

When analyzing large CAS, it is imperative to have a solid conceptual understanding of the relationship between critical, holistic features and such systems' properties. We have already described some of the features (eleven to be exact) so far. We would like to explain some additional less quantitative CAS features and properties. These additional features have been explained thoroughly in our previous and less quantitative book on Understanding CASs[41]. They are:

12. Degree of Contagion Effect (DCE)

When coupled, synchronized, coordinated, modulated, collaborative, cooperative, and synergistic local interactions among agents in a network exist, the CAS contagion effect

---

[40] Correlational analysis can be used in a system whenever there is uncertainty in how the cause and effect phenomena or Input and Output relationships work.
[41] **Shayan, S. A. (2019),**"Understanding Complex Adaptive Systems," Independent Publisher, Amazon.

occurs. CAS contagion can be a positive feature when there are multiple negative feedback and feedforward loops in the agent-agent structure since they can facilitate faster system adjustments towards equilibrium. It can be a negative feature when the system falls into a positive feedback and feedforward loop trap, and contagion causes a fast amplifying movement towards catastrophic and system breakdown. The contagion effect allows positive or negative information to spread across the system faster. *The more feedback and feedforward loops and Hamiltonian cycles in our agent-agent relationships, the higher the contagion effect in our CAS.* The existence of synchronization, cooperation, and modulation among CAS agents is a precondition for the contagion effect.

13. Degree of Collective Intelligence (DCI)

Most Complex Adaptive Systems and a particular collection of cognitive agents with no central controls, while following simple rules for interactions, can gain Collective Intelligence, which is sometimes known as Swarm Intelligence. It is a holistic, synergistic, distributive, and coordinated intelligence capable of showing dynamic and complex behavior, which would be difficult for an isolated agent to have individually. A more significant number of agents linked through simple rules and algorithms in CAS can result in higher average actualized/interactive information energy or Aep and holistic behavior that show resiliency, robustness resulting in dynamic equilibrium with a higher degree of collective intelligence.

14. Degree of Emergence and Innovation (DEI)

Emergence and innovation are holistic processes defined as the formation of order through synchronized, collective, and modulated local interactions of agents in a bottom-up and top-down manner giving rise to a sudden macrostructure through self-organizing and synergistic means. Emergence is an ad-hoc or discrete phenomenon, where localized behavior of agents aggregates into global behavior, disconnected from its origin. More heterogeneity or diversity of agents in CAS can lead to more flexibility and the possibility of observed emergent and new agent-agent interactions. A central force or process does not direct emergence. However, it results from a nonlinear interactive rule-based structure between agents, affecting the system at various levels and as a whole. The degree of Emergence and innovative behavior is higher when we have many cognitive agents interacting nonlinearly. Emergence over time leads to CAS behavior creating new structures, leading to more resilient, robust, and adaptable systems. In learning, we adapt to recognize an existing set of patterns, usually spelled out by the environment. In contrast, with emergence and innovation, we may suddenly view things differently. *The more complex a system becomes, the more emergent behavior in the short term and innovative dynamics in the long term it will have.*

15. Degree of Resilience (DRE)

Resilience describes a CAS's ability to persist and maintain its core functions or purpose in the presence of noise, disturbances, perturbation, stresses, or other changes in its inputs or environment. *The degree of resilience of a CAS is directly related to the degree of adaptability that it preserves. The degree of resilience is enhanced when we have locally distributed interacting dynamic controls. Therefore more central controls negatively impact the CAS's resilience and adaptability.*

16. Degree of Robustness (DRO)

Like resilience, robustness is the capability of a CAS to correct shocks, errors, noises, disturbances, or perturbations created in its inputs and structures and its ability to go back to its equilibrium state, but in the short term. *Therefore resilience is the long-term effect of robustness. The more robust a CAS is, the quicker it moves toward a state of equilibrium.* It means that CAS's robustness is a function of the time reached to equilibrium under shocks. Another way to describe robustness is a CAS's ability to maintain a particular behavior, trait, or characteristic regardless of changing environmental conditions. *The degree of robustness in a CAS is enhanced when we have locally distributed interacting and dynamic controls. More central controls negatively impact the CAS's robustness. More existing distributed controls between agents, such as feedback loop structures, lead to more CAS robustness, resulting in a quicker distribution of the effects of shock or noise throughout the CAS and a faster movement towards a state of equilibrium.*

The first eleven properties are fundamental, and the last five are general features that should be looked at in any CAS structure. These sixteen properties must jointly be reviewed and measured as best as possible to get a sense of the degree of complexity of a CAS.

Valuable common traits observed among all CAS are:

- They have a large number of interacting cognitive agents without central controls.
- Agents have dynamic physical or informational, one or two-way links and interactions.
- Agents have different impacts, weights, or influences to/from other agents.
- Agent interaction impacts are profound, simultaneous, and at all levels of the system.
- Agent interactions are non-linear.
- Agent interactions are primarily but not exclusively with immediate neighbors, and the influences are modulated, correlated with coupling and contagion effects.
- Agent interactions can negatively or positively feedback and feedforward onto themselves directly or after many intervening stages. We call this the recurrence effect.
- Agents collectively show cognitive, adaptive[42], and learning properties, making the system resilient, robust, innovative, and sustainable due to perturbation and shocks from the environment.
- Coevolving agents have mutual understanding and adaptation amongst one another
- Complex Adaptive Systems are open systems.
- System dynamics usually result in emergent, new, innovative, collective, and holistic behaviors.
- The system can filter out noise and error from input information due to its cognitive ability, coarse-graining, learning, and identifying correct patterns and regularities.
- Due to system openness and two-way interactions with the environment, CAS is always dynamic, continuous, and changing equilibrium.

---

[42] There is a distinction between being cognitive and adaptive. If a system is cognitive it will have some degree of adaptiveness. The more cognitive a system is the more adaptive it can become.

- Agent-agent correlational, cooperative, modulated, and contagion effects allow the system to have a history and memory. These systems evolve, and their past is co-responsible for the present and future behavior. There are nonlinear correlations between the past, present, and future states of the system. We define this as having memory or potential energy ingrained in the structure of the system.
- The competition will speed up learning and structural sophistication, and movement towards more complexity.
- Simply put, in all CAS, two essential features are always present; Diversity of agents and nonlinear internal rules between agents.

Examples of CAS include:
- ✓ International diplomacy and political systems
- ✓ International agricultural trades
- ✓ International banking transactions
- ✓ International digital currency trades and settlement
- ✓ International derivative market trades and settlements
- ✓ International bond market trades and settlements
- ✓ International stock market trades and settlements
- ✓ The human brain decision-making process

As a quantitative tool, the following table shows the measurement metrics for the sixteen defined parameters.

Complex Adaptive Systems
"Measurement Parameters"

Parameter	Code	Wt.	Measurement Metrics	Description
CAS Type	CT	-	PC, LC, IC, BC	As described in the section
Number of Agents	NOA	1	Low (0-3), Medium (3-6), High (6-9), and Very High (9-10)	Based on the number of agents or nodes, we allocate the numbers in the brackets. More NOA can cause more complexity.
Degree of Non-homogeneity of Agents	DNHA	3	None (0), Very Low (0-2), Low (2-4), Medium (4-6), High (6-8), and Very High (8-10)	Based on judgment and possible calculations ranging from None to Very High, we assign the number in the brackets. More DNHA can cause more complexity.
Average Number of Connections per Agents	ANCA	2	Low (0-3), Medium (3-6), High (6-9), and Very High (9-10)	Based on the number of connections per agent or node, we allocate the numbers in the brackets. More ANCA can cause more complexity.
Average Number of In Connections per Agent	ANICA	1	Low (0-3), Medium (3-6), High (6-9), and Very High (9-10)	We allocate the numbers in the brackets based on the number of In connections per agent or node. More ANICA can cause more complexity.
Average Number of Out Connections per Agent	ANOCA	1	Low (0-3), Medium (3-6), High (6-9), and Very High (9-10)	We allocate the numbers in the brackets based on the number of out connections per agent or node. More ANOCA can cause more complexity.

Parameter	Symbol	Value	Range	Description
Agent Coupling Strengths	ACS	2	Very Low (0-2), Low (2-4), Medium (4-6), High (6-8), and Very High (8-10)	Based on judgment and possible calculations ranging from Very Low to Very High, we assign the number in the brackets. More ACS can cause more complexity.
Number of Hamiltonian Cycles	NHC	3	None (0), Low (0-2), Medium (2-5), High (5-8) and Very High (8-10)	Based on judgment and possible calculations ranging from None to Very High, we assign the number in the brackets. More NHC can cause more complexity.
Average System Information Entropy	ASIP	-1	None (0), Very Low (0-2), Low (2-4), Medium (4-6), High (6-8), and Very High (8-10)	Based on judgment and possible calculations ranging from None to Very High, we assign the number in the brackets. More ASIP can cause complexity reduction.
Average System Information Energy	ASIG	2	None (0), Very Low (0-2), Low (2-4), Medium (4-6), High (6-8), and Very High (8-10)	Based on judgment and possible calculations ranging from None to Very High, we assign the number in the brackets. More ASIG can increase complexity.
System Dynamics	SD	2	Very Simple (0-3), Simple (3-6), Complex (6-8), Very Complex (8-10)	Based on judgment and possible equations determining the range from Very Simple to Very Complex, we assign the number in the brackets. More complex SD can cause more complexity.
Degree of Contagion Effect	DCE	3	None (0), Very Low (0-2), Low (2-4), Medium (4-6), High (6-8), and Very High (8-10)	Based on judgment and possible calculations ranging from None to Very High, we assign the number in the brackets. More DCE can cause more complexity.
Degree of Collective Intelligence	DCI	3	None (0), Very Low (0-2), Low (2-4), Medium (4-6), High (6-8), and Very High (8-10)	Based on judgment and possible calculations ranging from None to Very High, we assign the number in the brackets. More DCI is a sign of more complexity.
Degree of Emergence & Innovation	DEI	3	None (0), Very Low (0-2), Low (2-4), Medium (4-6), High (6-8), and Very High (8-10)	Based on judgment and possible calculations ranging from None to Very High, we assign the number in the brackets. More DEI can be a sign of more complexity.
Degree of Resilience	DRE	1	None (0), Very Low (0-2), Low (2-4), Medium (4-6), High (6-8), and Very High (8-10)	Based on judgment and possible calculations ranging from None to Very High, we assign the number in the brackets. More DRE can be a sign of more complexity.
Degree of Robustness	DRO	1	None (0), Very Low (0-2), Low (2-4), Medium (4-6), High (6-8), and Very High (8-10)	Based on judgment and possible calculations ranging from None to Very High, we assign the number in the brackets. More DRO can be a sign of more complexity.
Weighted Average Degree of CAS	WADC		None (0), Very Low (0 to 2), Low (2+ to 4), Medium (4+ to 6), High (6+ to 8), and Very High (8+ to 10)	Based on the weighted average score received for the 15 parameters (excluding CT), we assign the degree of CAS.

We can use various measurement tools or judgments to rank and describe each of the fifteen parameters defined in this table (we exclude CT because it is a qualitative measure) and predict the weighted average degree of Adaptive Complexity using the associated weights of a system.

We will try to analyze and use what we have learned so far by using several examples:

Example 10:

Let us assume that we have two top soccer teams playing against one another (each having 11 players) with different strengths and skills for the final European championship. The rule is to win in two 45 minute halves by having the highest goals scored. For the sake of simplicity, we will assume that both teams play a 4-3-3 style (one goalkeeper, four defenders, three midfielders, and three forwards). All 22 players and the one head referee[43] will have two-way time-varying semi- random interactions with one another with various degrees (in reality, depending on each team's strategy, clusters of players could be playing with one another, or one or two players can be the critical agents on the field for each team). Determine this system's internal network classification, attributes, and interaction coupling strength matrix. Next, using the measurement parameter table, determine the degree of complexity of the ball passing patterns created.

Answer 10

The system's graph is shown without coupling interactions, where team one players are shown as ●, team two players as ○ and the referee as ☺ ;

Semi-Closed System

For the system internal network classification, attributes, and the interaction coupling strength matrix, we have:

- **System Graph:** Shown above without the links (too many)
- **CAS Type:** Blended (Physical, Logical)
- **System Attributes:**
  ✓ NOA   23 agents (22 players plus 1 head referees)

---

[43] In such important games there are probably four assistant referees, two on the side lines and two on the goal lines, but for the sake of simplicity we will assume that these additional referees provide data for the head referee in the middle of the game to make decisions and only the head referee interacts with the players.

✓	DNHA	Non-homogeneous
✓	ANCA	529 links (23 x 23) with an avg. of 23 per agent (assumes an agent can learn or effect itself)
✓	ANICA	529 links with an avg. 23 per agent
✓	ANOCA	529 links with an avg. 23 per agent
✓	ACS	Defined as $c_{ij}(t)$ not shown above
✓	NHC	Many depending on different plays.
✓	ASIP, ASIG	Due to the semi-random assumption for the behavior of each player/agent, the average information entropy will be relatively large. The more structured strategy each team has, the lower the information entropy and higher information energy will be.
✓	SD	Very complex. We have a complex adaptive system with 23 agents and trigger parameters that can be represented by a vector with 23 elements of $T_1(t)$ to $T_{23}(t)$ and a 23 x 23 interaction coupling matrix as follows:

$$T(t) = \begin{pmatrix} T_1(t) \\ \vdots \\ T_{23}(t) \end{pmatrix}$$

$$C(t) = \begin{pmatrix} c_{11}(t) & \cdots & c_{1\,23}(t) \\ \vdots & \ddots & \vdots \\ c_{23\,1}(t) & \cdots & c_{23\,23}(t) \end{pmatrix}$$

In such a very Complex Adaptive System, the state of each player plus referee and the interaction coupling matrix can mutually and dynamically affect one another and change through time. It leads to a co-interacting, co-evolving, or adaptive behavior between agents and their respective interactions. Mathematically we can express the equations of the system as 46 sets of coupled nonlinear partial differential equations. The general mathematical form for this system of equations can be expressed as:

$$\frac{dT_i(t)}{dt} = A\left(c_{ij}(t), T_j(t)\right) \quad for\ i,j\ from\ 1\ to\ 23$$

$$\frac{dc_{ij}(t)}{dt} = B\left(c_{ij}(t), T_j(t)\right) \quad for\ i,j\ from\ 1\ to\ 23$$

Here, functions A and B can differ for each player/agent and depend on each team's strategy and the role that the head coaches have defined[44]. The functions for the system are semi-stochastic and are indeed non-linear, which are dependent on the state vector $T_j(t)$) and the interaction coupling strength matrix elements $c_{ij}(t)$.

- **Interaction Coupling Strength Matrix:**

$$C(t) = \begin{pmatrix} c_{11}(t) & \cdots & c_{1\,23}(t) \\ \vdots & \ddots & \vdots \\ c_{23\,1}(t) & \cdots & c_{23\,23}(t) \end{pmatrix}$$

The matrix can dynamically change through time, and the change is mathematically represented as:

$$\frac{dC(t)}{dt} = \begin{pmatrix} \frac{dc_{11}(t)}{dt} & \cdots & \frac{dc_{1\,23}(t)}{dt} \\ \vdots & \ddots & \vdots \\ \frac{dc_{23\,1}(t)}{dt} & \cdots & \frac{dc_{23\,23}(t)}{dt} \end{pmatrix}$$

Let us see how the Measurement Parameter table looks like:

Complex Adaptive System

---

[44] To be a bit more specific for each agent or player/referee, we have two equations for each player for a total of 46 sets of coupled nonlinear partial differential equations as follows:

$\frac{dT_1(t)}{dt} = A_1\left(c_{1j}(t), T_j(t)\right)$ for j from 1 to 23 for the **1st agent**

$\frac{dc_{1j}(t)}{dt} = B_1\left(c_{1j}(t), T_j(t)\right)$ for j from 1 to 23 for the **1st agent**

....

$\frac{dT_{23}(t)}{dt} = A_{23}\left(c_{23j}(t), T_j(t)\right)$ for j from 1 to 23 for the **23rd agent**

$\frac{dc_{23j}(t)}{dt} = B_{23}\left(c_{23j}(t), T_j(t)\right)$ for j from 1 to 23 for the **23rd agent**

European Final Championship
"Measurement Parameters"

Parameter	Code	Wt.	Measurement Metrics
CAS Type	CT	-	Blended (Physical, Logical)
Number of Agents	NOA	1	Very High (10)
Degree of Non-homogeneity of Agents	DNHA	3	Very High (10)
Average Number of Connections per Agents	ANCA	2	Very High (10)
Average Number of In Connections per Agent	ANICA	1	Very High (10)
Average Number of Out Connections per Agent	ANOCA	1	Very High (10)
Agent Coupling Strengths	ACS	2	Medium (5)
Number of Hamiltonian Cycles	NHC	3	Medium (5)
Average System Information Entropy	ASIP	-1	Medium (5)
Average System Information Energy	ASIG	2	Medium (5)
System Dynamics	SD	2	Very Complex (10)
Degree of Contagion Effect	DCE	3	High (7)
Degree of Collective Intelligence	DCI	3	Very High (9)
Degree of Emergence & Innovation	DEI	3	Very High (10)
Degree of Resilience	DRE	1	Very High (10)
Degree of Robustness	DRO	1	Very High (10)
Weighted Average Degree of CAS	WADC		Very High (8.4)

We have a very complex system with a WADC of 8.4. From such championship games, we expect no less. We could observe plenty of innovative and emerging behavior in a continuously dynamic setup. The game will also be resilient, robust to changes from 0-0 results (equal strengths continuously attacking and defending).

Example 11:

If in the previous question, the number of passes equal to x, the probability distribution of the number of passes for each player i in the 90-minute game was equal to $P_i(x)$ and was independent of the other players and followed the following distribution function with μ_i

and σ_i being player i historical average and standard deviation passing records:

$$P_i(x) = x_{i0} e^{-0.5\left(\frac{x-\mu_i}{\sigma_i}\right)^2} \quad \text{and} \quad \sum_{x=0}^{x=x} x_{i0} e^{-0.5\left(\frac{x-\mu_i}{\sigma_i}\right)^2} = 1$$

Find out the Average System Information Entropy or ASIP.

Answer 11

- **ASIP equals to:**

$$S_1 = -\sum_{x=0}^{x=x} (x_{10} e^{-0.5\left(\frac{x-\mu_i}{\sigma_i}\right)^2}) \ln(x_{10} e^{-0.5\left(\frac{x-\mu_i}{\sigma_i}\right)^2})$$

$$S_2 = -\sum_{x=0}^{x=x} (x_{20} e^{-0.5\left(\frac{x-\mu_i}{\sigma_i}\right)^2}) \ln(x_{20} e^{-0.5\left(\frac{x-\mu_i}{\sigma_i}\right)^2})$$

....

$$S_{23} = -\sum_{x=0}^{x=x} (x_{230} e^{-0.5\left(\frac{x-\mu_i}{\sigma_i}\right)^2}) \ln(x_{230} e^{-0.5\left(\frac{x-\mu_i}{\sigma_i}\right)^2})$$

$$ASIP = \frac{S_1 + S_2 + \cdots + S_{23}}{23}$$

Example 12:

If in example 10, the probability distribution of the number of balls passed for all players were independent but the same for all players with equal average μ and standard deviations σ for all or:

$$P(x) = x_0 e^{-0.5\left(\frac{x-\mu}{\sigma}\right)^2} \quad \text{and} \quad \sum_{x=0}^{x=x} x_0 e^{-0.5\left(\frac{x-\mu}{\sigma}\right)^2} = 1$$

Find out the Average System Information Entropy or ASIP.

Answer 12

- **ASIP, in this case, equals to:**

$$S_1 = -\sum_{x=0}^{x=x}(x_0 e^{-0.5\left(\frac{x-\mu}{\sigma}\right)^2})\ln(x_0 e^{-0.5\left(\frac{x-\mu}{\sigma}\right)^2})$$

$$S_2 = -\sum_{x=0}^{x=x}(x_0 e^{-0.5\left(\frac{x-\mu}{\sigma}\right)^2})\ln(x_0 e^{-0.5\left(\frac{x-\mu}{\sigma}\right)^2})$$

....

$$S_{23} = -\sum_{x=0}^{x=x}(x_0 e^{-0.5\left(\frac{x-\mu}{\sigma}\right)^2})\ln(x_0 e^{-0.5\left(\frac{x-\mu}{\sigma}\right)^2})$$

$$ASIP = \frac{S_1 + S_2 + \cdots + S_{23}}{23} = -\sum_{x=0}^{x=x}(x_0 e^{-0.5\left(\frac{x-\mu}{\sigma}\right)^2})\ln(x_0 e^{-0.5\left(\frac{x-\mu}{\sigma}\right)^2})$$

Example 13:

In example 11, as σ increases (standard deviation, volatility, or uncertainty in the number of passes the players make in a 90-minute game), what will happen to ASIP and ASIG.

Answer 13

As σ increases, $e^{-0.5\left(\frac{x-\mu}{\sigma}\right)^2}$ increases causing ASIP or information uncertainty to increase and ASIG to decrease. It means that as σ or volatility and uncertainty in the number of passes per player goes up, average system entropy goes up, and less structural certainty prevails for each player's passing behavior. The players would not have a structured game or strategy, and random plays prevail.

Example 14:

Using the analysis in the measurement parameter table, compare the degree of CAS for the following two complex adaptive systems:

- ✓ The movement patterns of 100,000 similar birds from east to west in the state of Minnesota. The birds have all the necessary space to move around freely, and they follow a straight line for 200 miles. While flying, each bird interacts with and follows five other birds in its immediate vicinity.
- ✓ One hundred thousand protestors are supporting Covid-19 third stimulus check walking down a major street in Washington DC. These are individuals from all walks of life and professions. They are all connected with cell phones and can communicate over social media. While walking, each person interacts and gets feedback from and provides feedforward to 7 other individuals. The Washington DC police need to analyze and predict crowd behavior to manage the possible chaotic and unpredictable behavior.

Compare the two systems and provide management recommendations regarding containing the CAS dynamics and preventing any unpredictable, unwanted emergent behavior.

Answer 14

Complex Adaptive System
100,000 Birds vs. Protestors
"Relative Measurement Parameters"

Parameter	Code	Wt.	Measurement Metrics for 100,000 birds	Measurement Metrics for 100,000 protestors
CAS Type	CT	-	Blended (Physical, Logical)	Blended (Physical, Logical)
Number of Agents	NOA	1	Very High (10)	Very High (10)
Degree of Non-homogeneity of Agents	DNHA	3	Low (3)	Very High (9)
Average Number of Connections per Agents	ANCA	2	Medium (5)	High (8)
Average Number of In Connections per Agent	ANICA	1	Medium (5)	High (8)
Average Number of Out Connections per Agent	ANOCA	1	Medium (5)	High (8)
Agent Coupling Strengths	ACS	2	Medium (4)	High (8)
Number of Hamiltonian Cycles	NHC	3	Medium (5)	High (7)
Average System Information Entropy	ASIP	-1	Medium (5)	Low (3)

Average System Information Energy	ASIG	2	Medium (5)	High (8)
System Dynamics	SD	2	Simple (4)	Complex (8)
Degree of Contagion Effect	DCE	3	Medium (4)	High (7)
Degree of Collective Intelligence	DCI	3	Medium (5)	High (8)
Degree of Emergence & Innovation	DEI	3	Low (3)	High (7)
Degree of Resilience	DRE	1	Medium (5)	High (8)
Degree of Robustness	DRO	1	Medium (5)	High (8)
Weighted Average Degree of CAS	WADC		Medium (4.5)	High (8.0)

The WADC for the bird movement is 4.5 (Medium) compared to the protestor movements being 8.0 (High). The bird movements do not seem to be posing any problems. On the other hand, the Washington DC police must be cautious and monitor the crowd movement. The degree of CAS for the protestors is reduced if the following actions are taken:

- ✓ Reduce the ANCA, ANICA, ANOCA, ACS, NHC, DCE, and DCI metrics by not allowing mobile phones or social media connections available in the protest path.
- ✓ Having the crowd be divided into smaller clusters with gaps in between in order to reduce the connections between them.
- ✓ Making sure that the path is taken as planned.

Example 15:

How do ASIG, ASIP, and information structural symmetry change when creating a language, writing a book, or a computer program?

Answer 15

In all three cases, ASIG increases, ASIP decreases, and the information structural symmetry decreases.

Example 16:

Using the measurement parameter table, compare the degree of CAS for the following two complex adaptive systems:

- ✓ Global Bitcoin trading network (a form of crypto-currency) with no central regulatory mechanism. This digital currency is being traded through a secure Blockchain platform with the global trading members or agents' self-regulatory mechanism. It is a highly inter-related 7/24 competitive trading environment.
- ✓ The initial mental process to create classical music, such as Bach's double violin concerto. It requires many complex interactions between brain neurons (1 hundred billion where each has an average of 7000 connection to other neurons) as agents, maintaining existing memory and all physical senses.

Answer 16

The results are shown in the following table:

Complex Adaptive System
Global Bitcoin Trading vs. Creation of a Classical Music
"Relative Measurement Parameters"

Parameter	Code	Wt.	Global Bitcoin Trading	Classical Music Creation
CAS Type	CT	-	Blended (Logical, Informational)	Blended (Physical, Logical, Info.)
Number of Agents	NOA	1	Very High (10)	Very High (10)
Degree of Non-homogeneity of Agents	DNHA	3	Very High (10)	Very High (10)
Average Number of Connections per Agents	ANCA	2	High (8)	Very High (10)
Average Number of In Connections per Agent	ANICA	1	High (8)	Very High (10)
Average Number of Out Connections per Agent	ANOCA	1	High (8)	Very High (10)
Agent Coupling Strengths	ACS	2	High (8)	Very High (10)
Number of Hamiltonian Cycles	NHC	3	Very High (10)	Very High (10)
Average System Information Entropy	ASIP	-1	Medium (5)	Very High (9)
Average System Information Energy	ASIG	2	Medium (5)	Very Low (2)
System Dynamics	SD	2	Very Complex (10)	Very Complex (10)
Degree of Contagion Effect	DCE	3	Very High (10)	Very High (10)
Degree of Collective Intelligence	DCI	3	Medium (5)	Very High (10)
Degree of Emergence & Innovation	DEI	3	Medium (5)	Very High (10)

Degree of Resilience	DRE	1		High (7)	Very High (10)
Degree of Robustness	DRO	1		High (7)	Very High (10)
Weighted Average Degree of CAS	**WADC**			**High (8.0)**	**Very High (9.4)**

The WADC for Global Bitcoin trading is high and equal to 8.0. it is a 7/24, very interactive, and continuous trading platform without any central regulatory authority restrictions. It is a highly competitive and dynamic trading environment. On the other hand, the mental process to create beautiful classical music such as Bach's double violin concerto from scratch has a WADC equal to 9.4, which is even higher. It is a highly creative, adaptive, co-evolving, and interactive process that billions of neurons dynamically go through. It leads to many emerging and possible innovative creations. The main reasons for a higher degree of CAS are higher ANCA, ANICA, ANOCA, DCI, DEI, DRE, and DRO.

To manage the complexity levels of an adaptive system, we should first analyze the Measurement Parameters table and calculate WADC. Next, depending on the intention to increase or decrease complexity, an appropriate strategy can be adopted. The table below is based on the intention to reduce WADC of 10 in a CAS.

Complex Adaptive Systems
"How to Reduce Complexity through Measurement Parameters"

Parameter	Code	Wt.	Measurement Metrics	How to reduce WADC
CAS Type	CT	-	PC, LC, IC, BC	If possible, break the blended nature into distinct types.
Number of Agents	NOA	1	Very High (10)	Reduce NOA by eliminating unnecessary agents, breaking up the total number of agents into many smaller homogeneous clusters, and minimizing cluster-cluster interactions.
Degree of Non-homogeneity of Agents	DNHA	3	Very High (10)	Reduce DNHA by separating agents into homogeneous clusters as separate active units and minimizing the cluster-cluster interactions.
Average Number of Connections per Agents	ANCA	2	Very High (10)	If possible, reduce ANCA by eliminating links and connections.
Average Number of In Connections per Agent	ANICA	1	Very High (10)	If possible, reduce ANICA by eliminating in connections.
Average Number of Out Connections per Agent	ANOCA	1	Very High (10)	If possible, reduce ANOCA by eliminating out connections.
Agent Coupling Strengths	ACS	2	Very High (10)	If possible, reduce ACS by breaking up some internal structures or agent-agent feedback/feedforward interactions.

Number of Hamiltonian Cycles	NHC	3	Very High (10)	If possible, reduce NHC by breaking the NHC loops and internal structures or agent-agent feedback/feedforward interactions.
Average System Information Entropy	ASIP	-1	Very High (10)	If possible, increase ASIP by breaking up some internal structures or agent-agent feedback/feedforward interactions.
Average System Information Energy	ASIG	2	Very High (10)	If possible, reduce ASIG by breaking up some internal structures or agent-agent feedback/feedforward interactions.
System Dynamics	SD	2	Very Complex (10)	If possible, reduce or eliminate the coupling features of the total system of equations. Make the system of equations independent from one another and less complex.
Degree of Contagion Effect	DCE	3	Very High (10)	If possible, reduce DCE by reducing ANCA, ANICA, ANOCA, NHC and breaking up some internal structures or agent-agent feedback/feedforward interactions.
Degree of Collective Intelligence	DCI	3	Very High (10)	Reduce DCI by reducing ANCA, ANICA, ANOCA, ACS, NHC, DCE, breaking up the total number of agents into many smaller similar clusters, and minimizing cluster-cluster interactions. Also, reduce the number of NHCs and break up some internal structures or agent-agent feedback/feedforward interactions.
Degree of Emergence & Innovation	DEI	3	Very High (10)	Reduce DEI by reducing ANCA, ANICA, ANOCA, ACS, NHC, DCE, breaking up the total number of agents into many smaller similar clusters, and minimizing cluster-cluster interactions. Also, reduce the number of NHCs and break up some internal structures or agent-agent feedback/feedforward interactions.
Degree of Resilience	DRE	1	Very High (10)	Reduce DRE by reducing ANCA, ANICA, ANOCA, ACS, NHC, DCE, and breaking up some internal structures or agent-agent feedback/feedforward interactions.
Degree of Robustness	DRO	1	Very High (10)	Reduce DRO by reducing ANCA, ANICA, ANOCA, ACS, NHC, DCE, and breaking up some internal structures or agent-agent feedback/feedforward interactions
Weighted Average Degree of CAS	WADC		Very High (10)	

There are many instances at which a given CAS's degree of complexity can be reduced by one action, such as separating agents into homogeneous clusters as separate active units and minimizing the cluster-cluster interactions alone. Sometimes we might need to take several actions. It all depends on the internal dynamics of a CAS. The robustness and resilient capabilities of a CAS to correct shocks, errors, noises, disturbances, or perturbations and the ability to go back to a new equilibrium or emergent state can be used to move it from one state of equilibrium to another. The more robust a CAS is, the faster the equilibrium of a system is reached. Injecting controlled noise, perturbation, or shock can also be considered as an effective CAS management tool when a new equilibrium state is aimed. The injected shock or noise should not be above certain critical levels (which is sometimes difficult to know) because it can push the system into a state of chaos, unpredictability, instability, or self-organized criticality[45]. ***In***

---

[45] **Self-Organized Criticality**; refers to the notion that there can be situations where a CAS can have critical points of triggering and reshaping that will be reached

general, a CAS must always stay in a state of changing equilibrium to be adaptable, resilient, and robust.

At this point, we must expand the concept of Swarm Intelligence[46] (see the Definition section). We have already briefly explained this concept in our previous definition of the degree of Collective Intelligence or DCI. Due to its importance, we will spend more time on the concept.

Swarm Intelligence Structure (SIS) is a holistic, synergistic, non-centralized, distributive, and coordinated intelligence structure observed when we have multiple cognitive agents with no central controls, grouped while following simple rules and algorithms to achieve a stable, dynamically equilibrated goal. These goals can be the survival of the species under external threat, traffic reduction, or finding an equilibrium state through the Least Action Principle's utilization, which deals with the general methodology of finding the optimum and equilibrium states or function of a multi-dimensional many-body system.

The dynamics of Swarm Intelligence is based on a simple CAS structure which we abbreviate as **AIAR**, with the following capabilities:

- Each cognitive agent communicates, interacts, and is aware of its surroundings – **Awareness.**
- Each cognitive agent is autonomous and makes independent decisions but follows an algorithm – **Independence.**
- Each cognitive agent is aware of what it can do – **Autonomy.**
- Cognitive agents can be added, eliminated, and replaced with no effect on the system – **Resilience or Robustness.**

Swarm intelligence can solve complex problems in cases where it is difficult for an isolated agent to solve. Swarm Intelligence Structures have recently been used in small robots in large numbers using strategies inspired from nature similar to fish, locust, bird, bee, and ant colonies. Research has shown that SIS in decision-making processes improves decision accuracy by up to forty percent with more innovative, emergent, and robust solutions.

---

spontaneously throughout its dynamics. This is a point where the CAS equilibrium can become unstable, highly sensitive, and responsive to shocks, errors, or changes resulting in reduced levels of resiliency and robustness. At this stage the system has sensitive dependence on its initial conditions. Cases such as financial bubbles and meltdowns, social and political revolutions, avalanches, and ferromagnetic phase changes fall under such behavior.

[46] Also see, **Shayan, S. A. (2019)**,"Understanding Complex Adaptive Systems," Independent Publisher, Amazon.

If we apply the measurement parameter table to general Swarm Intelligence Structures, we get:

Complex Adaptive Systems
"Swarm Intelligence Structure Measurement Parameters"

Parameter	Code	Wt.	Measurement Metrics
CAS Type	CT	-	PC, LC, IC, BC
Number of Agents	NOA	1	Very High (10)
Degree of Non-homogeneity of Agents	DNHA	3	Medium (5)
Average Number of Connections per Agents	ANCA	2	Medium (5)
Average Number of In Connections per Agent	ANICA	1	Medium (5)
Average Number of Out Connections per Agent	ANOCA	1	Medium (5)
Agent Coupling Strengths	ACS	2	Medium (5)
Number of Hamiltonian Cycles	NHC	3	Medium (5)
Average System Information Entropy	ASIP	-1	Medium (5)
Average System Information Energy	ASIG	2	Medium (5)
System Dynamics	SD	2	Complex (6)
Degree of Contagion Effect	DCE	3	Medium (5)
Degree of Collective Intelligence	DCI	3	High (7)
Degree of Emergence & Innovation	DEI	3	High (7)
Degree of Resilience	DRE	1	High (7)
Degree of Robustness	DRO	1	High (7)
Weighted Average Degree of CAS	WADC		Medium (5.9)

An important and related concept to SIS is "Herd Mentality Phenomena," or HMP. It occurs when most agents are heavily influenced or follow one or a limited number of agents and mimic their behavior. Almost all agents become followers without thinking and lose their autonomy and independence. ***MHP behavior is a holistic, non-synergistic, centralized, non-distributive, non-cognitive, and non-coordinated structure led by one or few cognitive central agents, followed by most other agents without thinking or independence. This type of behavior can lead to positive or negative results depending on the lead agent's objectives.*** In HMP, agents lack **Awareness,**

Independence, and **Autonomy** hence **Resilience** and **Robustness**. The goals are not generally to find an equilibrium state and the ***Least Action Principle*** is not at work. HMP is an example of extreme correlational, emotional, or contagion behavior. One or a handful of agents can be manipulated to direct the agents' group towards where they intend it to go. It usually happens at times of high excitement, impaired cognition, confusion, and lack of direction. Cases such as financial bubble creations and collapses, street riot eruptions, and volatile international political behaviors leading to military conflicts can be categorized as HMP behaviors.

The following table compares general Swarm Intelligence Structures and Herd Mentality Phenomena:

Complex Adaptive System
Swarm Intelligence Structure vs. Herd Mentality Phenomena "Relative Measurement Parameters"

Parameter	Code	Wt.	Swarm Intelligence Structure	Herd Mentality Phenomena
CAS Type	CT	-	PC, LC, IC, BC	Blended (Physical, Logical, Info.)
Number of Agents	NOA	1	Very High (10)	Very High (10)
Degree of Non-homogeneity of Agents	DNHA	3	Medium (5)	Low (2)
Average Number of Connections per Agents	ANCA	2	Medium (5)	Low (2)
Average Number of In Connections per Agent	ANICA	1	Medium (5)	Low (2)
Average Number of Out Connections per Agent	ANOCA	1	Medium (5)	Low (2)
Agent Coupling Strengths	ACS	2	Medium (5)	Very High (9)
Number of Hamiltonian Cycles	NHC	3	Medium (5)	Low (2)
Average System Information Entropy	ASIP	-1	Medium (5)	Very Low (2)
Average System Information Energy	ASIG	2	Medium (5)	Medium (5)
System Dynamics	SD	2	Complex (6)	Simple (3)
Degree of Contagion Effect	DCE	3	Medium (5)	High (7)
Degree of Collective Intelligence	DCI	3	High (7)	Very Low (2)
Degree of Emergence & Innovation	DEI	3	High (7)	Very Low (2)
Degree of Resilience	DRE	1	High (7)	Very Low (2)
Degree of Robustness	DRO	1	High (7)	Very Low (2)
Weighted Average Degree of CAS	WADC		Medium (5.9)	Low (3.7)

As it can be seen, WADC for SIS is 5.9 and for HMP is 3.7 or close to 2.2 points lower. The only common parameter for both was the very high number of agents. The more synergistic, non-centralized, distributive, and coordinated intelligence structure with many SIS agents has led to the higher WADC value.

The possible advantages of observing complex adaptive emergent behavior can increase as Non-homogeneity of agents (DNHA) increases in SIS.

When we deal with complex management, economic, financial, social, political, military, or cultural decisions, it is advisable to use a SIS approach. A collection of non-homogeneous cognitive agents or individuals in committees, councils, or groups would be very effective with clarified rules and procedures to follow. The possibilities of observed errors and risks in decision-making procedures are reduced due to cognitive agents' self-correction through this mechanism. *SIS results in additional memory, order, corrections, emergent behavior, and complexity in the system.*

In all Swarm Intelligence Structures, the challenge is the determination of proper agent-agent rules and algorithms. The optimum structures are created to minimize the potential errors and risks for attaining the intended goals and defined objectives. If optimum objectives are not determined, the effects could be as inefficient as a failing behavior of a HMP in place.

Synchronization in CAS

Synchronization[47] is a phenomenon that occurs when two or many interacting bodies in a system, having certain critical coupling threshold levels, will coordination their behavior and operate in unison. A System that behaves with all parts in synchrony is said to be synchronous.

Due to agent-agent coupling features, synchronization is observed in all CAS. Any CAS that attains high ANCA, ANICA, ANOCA levels, and critical coupling threshold levels of ACS (agent coupling strengths) will possess the Synchronization capability, with lower ASIP and structural symmetries and higher ASIG. It means that if there is any

---

[47] Also see, **Manrubia, C. S., Mikhailov, A. S., and Zanette, D. H. (2004).** "Emergence of Dynamical Order, Synchronization Phenomena in Complex Systems." World Scientific Lecture Notes in Complex Systems – Vol. 2, Singapore.

variation or statistical change in any or multiple agents, the variability could end up synchronizing and propagating to a cluster or even the whole system. The system could emerge and move into a new equilibrium, resiliency, and robustness through this process. In the case of global synchronization of a CAS, an abrupt transition from unsynchronized to full synchronization occurs when the coupling strength exceeds the critical threshold levels. The threshold coupling level is unique for each CAS, and it is dependent on the structure and dynamics of the system. Very high coupling levels between agents could cause the system to collapse. If the coupling strengths are below the threshold levels, synchronization will not occur. In such cases, a shock or variability on any agent will not cause synchronization and dissipate or result in random fluctuations in the system. Synchronization in CAS could be caused by an external perturbation or a consequence of the interaction between cognitive agents.

Synchronization in CAS is an emergent property, and to analyze and measure it, advanced statistical and correlational approaches are required. The main challenge in any CAS synchronization analysis is determining the critical coupling threshold levels for each agent, existing clusters, and the whole system.

Examples of synchronization phenomena include:

- ✓ People clapping hands together at the end of an opera
- ✓ Fireflies lighting on a tree
- ✓ Electric excitation wave propagating through the heart
- ✓ Crowd walking or cars driving on a bridge

A society is a large CAS with active agents. The agents are organized into coordinated groups. Through joint and coupled efforts, a society achieves synergistic goals that are out of reach for its agents. Different collective tasks can be exercised if the proper structures are in place, depending on the interactions and degree of synchronization. Synchronization can play a critical role in resolving social issues.

Synchronization is possible in CAS due to the coupling features of the agents. It happens at the agent-agent levels when the critical coupling threshold levels have reached, expanding to either cluster or the whole system levels. Synchronization can be a driving force behind the adaptability, emergence, robustness, and resiliency of a CAS.

As stated before, In all Swarm Intelligence Structures, the challenge is determining proper agent-agent coupling rules and

algorithms. If these coupling rules and algorithms are set, such that the Swarm structure attains the required critical coupling threshold levels, a Synchronized Swarm Intelligence Structure (SSIS) could be observed. Such a CAS will be very adaptable, robust, and resilient. Current research on coordinated small drones and robots utilizing Artificial Intelligence is dealing with these concepts.

Questions

Question 1. Provide two examples of nonlinear behavior of n cognitive agents interacting.

Answer 1:
When two Hockey teams are competing, or members of an international NGO interact with several villages to reduce the poverty levels. These are both highly nonlinear cognitive interactions. These interactions are dynamic with possible emergent and innovative results.

Question 2. How would ASIG, ASIP, and structural symmetries be affected in the following systems?

- Classical music creation
- Painting by an artist
- Building a new house
- Writing a poem
- Two cars crashing

Answer 2:
The answer is shown in the following table

Function	ASIG	ASIP	Symmetry
Classical music creation	Increases	Decreases	Decreases
Painting by an artist	Increases	Decreases	Decreases
Building a new house	Increases	Decreases	Decreases
Writing a poem	Increases	Decreases	Decreases
Two cars crashing	Decrease	Increase	Increase

Question 3. Assume that we have a system of electric power generators set up in a network structure shown below. It is a closed system. All generator-generator (agent-agent) interactions are different and create a final output equal to a function related to the generator-generator interactions' linear cumulative effect. Generators and electric link flows are fixed and deterministic. What is the system's internal network classification, attributes, and interaction matrix?

Closed System

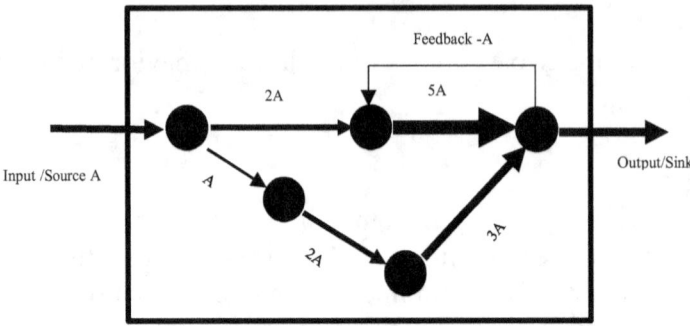

Answer 3:

We have,

- **System Graph:** Shown above
- **CAS Type:** Physical
- **System Attributes:**
  - ✓ NOA — 5 agents
  - ✓ DNHA — Non-homogeneous
  - ✓ ANCA — 8 links with an avg. 1.6 per agent including the source and sink
  - ✓ ANICA — 7 links with an avg. 1.4 per agent
  - ✓ ANOCA — 7 links with an avg. 1.4 per agent
  - ✓ ACS — Shown on the graph (A, 2A, 3A, 5A)
  - ✓ NHC — 1
  - ✓ ASIP — 0 with no variations in node behaviors
  - ✓ SD — Shown on the structure

- **Input/Source:** A
- **Output/Sink Result:** Agent5 output $3A + 5A - A = 7A$
- **Trigger Parameter Vector:**

$$T(t) = \begin{pmatrix} T_1 = 3A \\ T_2 = 2A \\ T_3 = 5A \\ T_4 = 3A \\ T_5 = 5A + 3A - A = 7A \end{pmatrix}$$

$$\frac{dT}{dt} = \begin{pmatrix} 0 \\ 0 \\ 0 \\ 0 \\ 0 \end{pmatrix}$$

- **Interaction Coupling Strength Matrix:**

$$C(t) = \begin{pmatrix} 0 & 1A & 2A & 0 & 0 \\ 0 & 0 & 0 & 2A & 0 \\ 0 & 0 & 0 & 0 & 5A \\ 0 & 0 & 0 & 0 & 3A \\ 0 & 0 & -1A & 0 & 8A \end{pmatrix}$$

$$\frac{dC(t)}{dt} = \begin{pmatrix} 0 & 0 & 0 & 0 & 0 \\ 0 & 0 & 0 & 0 & 0 \\ 0 & 0 & 0 & 0 & 0 \\ 0 & 0 & 0 & 0 & 0 \\ 0 & 0 & 0 & 0 & 0 \end{pmatrix}$$

Question 4. Assume that we have the same electric power generator system setup with nonlinearity in the previous example's internal structure, as shown below. Let us assume that interactions create a final output equal to a function related to a linear cumulative effect (due to nonlinearity, this will be considered as CAS). Generators and their links are all fixed and deterministic. Determine the system's internal network classification, attributes, and interaction coupling strength matrix.

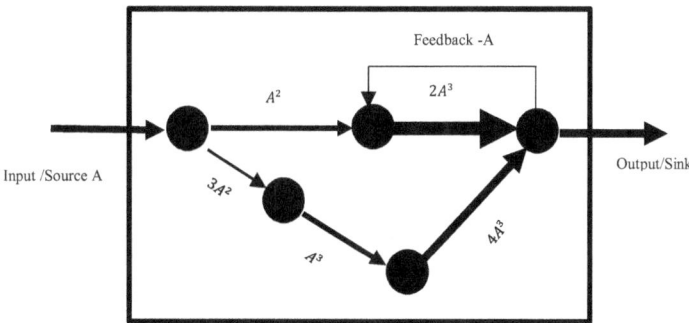

Answer 4:
For the system internal generator-generator (agent-agent) network classification, attributes and the interaction coupling strength matrix we have:

- **System Graph:** Shown above
- **CAS Type:** Physical
- **System Attributes:**

- ✓ NOA   5 Agents
- ✓ DNHA   Non-homogeneous
- ✓ ANCA   8 links with an avg. 1.6 per agent including the source and sink
- ✓ ANICA   7 links with an avg. 1.4 per agent
- ✓ ANOCA   7 links with an avg. 1.4 per agent
- ✓ ACS   Shown on the graph
- ✓ NHC   1
- ✓ ASIP   0 with no variations in node behaviors
- ✓ SD   Shown on the structure

- **Input/Source:**   A
- **Output/Sink Result:** Agent5 output $4A^3 + 2A^3 - A = 6A^3 - A$
- **Trigger Parameter Vector:**

$$T(t) = \begin{pmatrix} T_1 = 3A^2 + A^2 = 4A^2 \\ T_2 = A^3 \\ T_3 = 2A^3 \\ T_4 = 4A^3 \\ T_5 = 4A^3 + 2A^3 - A = 6A^3 - A \end{pmatrix}$$

$$\frac{dT(t)}{dt} = \begin{pmatrix} 0 \\ 0 \\ 0 \\ 0 \\ 0 \end{pmatrix}$$

- **Interaction Coupling Strength Matrix:**

$$C(t) = \begin{pmatrix} 0 & 3A^2 & A^2 & 0 & 0 \\ 0 & 0 & 0 & A^3 & 0 \\ 0 & 0 & 0 & 0 & 2A^3 \\ 0 & 0 & 0 & 0 & 4A^3 \\ 0 & 0 & -A & 0 & 6A^3 \end{pmatrix}$$

$$\frac{dC(t)}{dt} = \begin{pmatrix} 0 & 0 & 0 & 0 & 0 \\ 0 & 0 & 0 & 0 & 0 \\ 0 & 0 & 0 & 0 & 0 \\ 0 & 0 & 0 & 0 & 0 \\ 0 & 0 & 0 & 0 & 0 \end{pmatrix}$$

Question 5. If in the previous question, the probability distribution for the output source for each generator (agent) was independent and varied through time for each state c *(the price of gas to generate electricity)* as follows:

$$A(c) = e^{-ct} \quad \text{and} \quad \sum_{c=1}^{c=c} e^{-ct} = 1$$

Find out the generator Trigger Parameters, Interaction Parameter Coupling Strength Matrix, their changes through time, and the Average System Information Entropy (uncertainty) or ASIP.

Answer 5:
We get,

- **Trigger Parameter Vector:**

$$\boldsymbol{T}(t) = \begin{pmatrix} T_1 = 4e^{-2ct} \\ T_2 = e^{-3ct} \\ T_3 = 2e^{-3ct} \\ T_4 = 4e^{-3ct} \\ T_5 = 6e^{-3ct} - e^{-ct} \end{pmatrix}$$

$$\frac{d\boldsymbol{T}(t)}{dt} = \begin{pmatrix} -8ce^{-2ct} \\ -3ce^{-3ct} \\ -6ce^{-3ct} \\ -12ce^{-3ct} \\ -18ce^{-3ct} + ce^{-ct} \end{pmatrix}$$

- **Interaction Coupling Strength Matrix:**

$$\boldsymbol{C}(t) = \begin{pmatrix} 0 & 3e^{-2ct} & e^{-2ct} & 0 & 0 \\ 0 & 0 & 0 & e^{-3ct} & 0 \\ 0 & 0 & 0 & 0 & 2e^{-3ct} \\ 0 & 0 & 0 & 0 & 4e^{-3ct} \\ 0 & 0 & -e^{-ct} & 0 & 6e^{-3ct} \end{pmatrix}$$

$$\frac{d\boldsymbol{C}(t)}{dt} = \begin{pmatrix} 0 & -6ce^{-2ct} & -2ce^{-2ct} & 0 & 0 \\ 0 & 0 & 0 & -3ce^{-3ct} & 0 \\ 0 & 0 & 0 & 0 & -6ce^{-3ct} \\ 0 & 0 & 0 & 0 & -12ce^{-3ct} \\ 0 & 0 & ce^{-ct} & 0 & -18ce^{-3ct} \end{pmatrix}$$

- **ASIP equals to:**

$$S_1 = -\sum_{c=1}^{c=c} e^{-ct} \ln(e^{-ct}) = \sum_{c=1}^{c=c} cte^{-ct}$$

$$S_2 = -\sum_{c=1}^{c=c} e^{-ct} \ln(e^{-ct}) = \sum_{c=1}^{c=c} cte^{-ct}$$

$$S_3 = -\sum_{c=1}^{c=c} e^{-ct} \ln(e^{-ct}) = \sum_{c=1}^{c=c} cte^{-ct}$$

$$S_4 = -\sum_{c=1}^{c=c} e^{-ct} \ln(e^{-ct}) = \sum_{c=1}^{c=c} cte^{-ct}$$

$$S_5 = -\sum_{c=1}^{c=c} e^{-ct} \ln(e^{-ct}) = \sum_{c=1}^{c=c} cte^{-ct}$$

$$ASIP = \frac{S_1 + S_2 + S_3 + S_4 + S_5}{5} = \sum_{c=1}^{c=c} cte^{-ct}$$

Question 6. In the previous question, find out the agent Trigger Parameters, Interaction Parameter Strength Matrix, the changes through time, and the Average System Information Entropy or ASIP as t becomes very large or goes to infinity (this system can be considered as CAS).

Answer 6:
We get,

- **Trigger Parameter Vector:**

$$\lim_{t \to \infty} \boldsymbol{T}(t) = \begin{pmatrix} T_1 = 4e^{-2ct} = 0 \\ T_2 = e^{-3ct} = 0 \\ T_3 = 2e^{-3ct} = 0 \\ T_4 = 4e^{-3ct} = 0 \\ T_5 = 6e^{-3ct} - e^{-ct} = 0 \end{pmatrix}$$

$$\lim_{t \to \infty} \frac{d\mathbf{T}(t)}{dt} = \begin{pmatrix} 0 \\ 0 \\ 0 \\ 0 \\ 0 \end{pmatrix}$$

- **Interaction Coupling Strength Matrix:**

$$\lim_{t \to \infty} \mathbf{C}(t) = \begin{pmatrix} 0 & 0 & 0 & 0 & 0 \\ 0 & 0 & 0 & 0 & 0 \\ 0 & 0 & 0 & 0 & 0 \\ 0 & 0 & 0 & 0 & 0 \\ 0 & 0 & 0 & 0 & 0 \end{pmatrix}$$

$$\lim_{t \to \infty} \frac{d\mathbf{C}(t)}{dt} = \begin{pmatrix} 0 & 0 & 0 & 0 & 0 \\ 0 & 0 & 0 & 0 & 0 \\ 0 & 0 & 0 & 0 & 0 \\ 0 & 0 & 0 & 0 & 0 \\ 0 & 0 & 0 & 0 & 0 \end{pmatrix}$$

- **ASIP equals to:**

$$\lim_{t \to \infty} S_1 = 0$$

$$\lim_{t \to \infty} S_2 = 0$$

$$\lim_{t \to \infty} S_3 = 0$$

$$\lim_{t \to \infty} S_4 = 0$$

$$\lim_{t \to \infty} S_5 = 0$$

$$\lim_{t \to \infty} ASIP = \lim_{t \to \infty} \frac{S_1 + S_2 + S_3 + S_4 + S_5}{5} = 0$$

It means that as time goes on, we reach towards more output certainty (zero ASIP) for each generator (agent), independent of the price of the gas.

Question 7. If t goes to zero for the previous question, find out the agent Trigger Parameters, Interaction Coupling Strength Matrix, the changes through time, and the Average System Information Entropy or ASIP.

Answer 7:
We get,

- **Trigger Parameter Vector:**

$$\lim_{t \to 0} \mathbf{T}(t) = \begin{pmatrix} T_1 = 4 \\ T_2 = 1 \\ T_3 = 2 \\ T_4 = 4 \\ T_5 = 5 \end{pmatrix}$$

$$\lim_{t \to 0} \frac{d\mathbf{T}(t)}{dt} = \lim_{t \to 0} \begin{pmatrix} -8c \\ -3c \\ -6c \\ -12c \\ -17c \end{pmatrix}$$

- **Interaction Coupling Strength Matrix:**

$$\lim_{t \to 0} \mathbf{C}(t) = \begin{pmatrix} 0 & 3 & 1 & 0 & 0 \\ 0 & 0 & 0 & 1 & 0 \\ 0 & 0 & 0 & 0 & 2 \\ 0 & 0 & 1 & 0 & 4 \\ 0 & 0 & -1 & 0 & 6 \end{pmatrix}$$

$$\lim_{t \to 0} \frac{d\mathbf{C}(t)}{dt} = \begin{pmatrix} 0 & -6c & -2c & 0 & 0 \\ 0 & 0 & 0 & -3c & 0 \\ 0 & 0 & 0 & 0 & -6c \\ 0 & 0 & 0 & 0 & -12c \\ 0 & 0 & c & 0 & -18c \end{pmatrix}$$

- **ASIP equals to:**

$$\lim_{t \to 0} S_1 = 0$$

$$\lim_{t \to 0} S_2 = 0$$

$$\lim_{t \to 0} S_3 = 0$$

$$\lim_{t \to 0} S_4 = 0$$

$$\lim_{t \to 0} S_5 = 0$$

$$\lim_{t \to 0} ASIP = \lim_{t \to 0} \frac{S_1 + S_2 + S_3 + S_4 + S_5}{5} = 0$$

It also means that initially, the network had output certainty (zero ASIP) for each generator (agent), independent of the gas's price.

Question 8. Compare three states of *t* going from zero to interim and infinity for the previous example. Assuming that no change is made to the CAS agents or their links, determine which state has higher and lower average system information entropy and energy.

Answer 8
We have the following table:

CAS parameter	$t \to 0$	$t \to t$	$t \to \infty$
ASIP	zero	Maximum	zero
ASIG	Highest	Minimum	Highest

Initially, the generator's output uncertainty in the network is the lowest (ASIP equals zero), and the average system information energy or ASIG is the highest. As time goes on, the output uncertainty increases and ASIP is maximized, causing the ASIG to be minimized. In the long run, the system will go back to zero output uncertainty and having the highest ASIG. There is a point in time ($t = t$) where the output uncertainty or ASIP is the largest or maximized, and ASIG is the lowest or minimized.

Question 9. Assume that we have a closed system that generates energy as its output. It has an input that can vary with a time equal to *x(t)*. We also have three energy-generating units (agents or nodes), each making certain energy transformations shown in the system graph below. The first node creates two outputs; one equals to the original input *x(t)* times a constant *a* and the second as a derivative of the initial input with regards to time or $\frac{dx(t)}{dt}$. The second node takes the $\frac{dx(t)}{dt}$ input and generates an output as a second derivative with regards to time or $\frac{d^2x(t)}{dt^2}$ times a constant b. The constants a and b are system factors that can be determined by design. Let us assume that interaction between the first and the second nodes, when added linearly, create a final output equal to $x^2(t)$ times a constant c. For various values of a, b, and c, we will get different system dynamic behaviors. Due to the nonlinearity of interactions, this system will be considered CAS. The number of energy generators and their links are all fixed and do not change. Determine the system's internal network classification, attributes, and interaction coupling strength matrix.

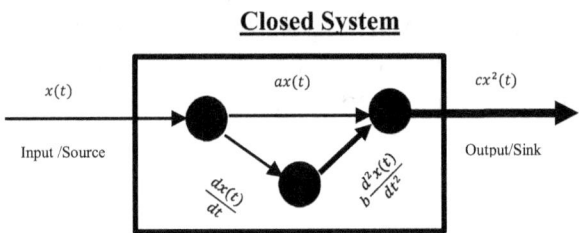

Answer 9

For the system internal generator-generator (agent-agent) network classification, attributes and the interaction coupling strength matrix we have:

- **System Graph:** Shown above
- **CAS Type:** Physical
- **System Attributes:**
  - ✓ NOA   3 Agents
  - ✓ DNHA   Non-homogeneous
  - ✓ ANCA   5 links with an avg. 1.7 per agent including the source and sink
  - ✓ ANICA   4 links with an avg. 1.3 per agent
  - ✓ ANOCA   4 links with an avg. 1.3 per agent
  - ✓ ACS   Shown on the graph
  - ✓ NHC   0
  - ✓ ASIP   0 with no variations in node behaviors
  - ✓ SD $b\frac{d^2x(t)}{dt^2} = cx^2(t) - ax(t)$ this is a 2nd order, 2nd degree, Continuous, Non-linear, Non-autonomous, deterministic equation of the system. Parameters a, b, and c are constants.
- **Input/Source:** $x(t)$

- **Output/Sink Result:** Agent3 output $cx^2(t)$
- **Trigger Parameter Vector:**

$$T(t) = \begin{pmatrix} T_1 = \dfrac{dx(t)}{dt} + ax(t) \\ T_2 = b\dfrac{d^2x(t)}{dt^2} \\ T_3 = b\dfrac{d^2x(t)}{dt^2} + ax(t) = cx^2(t) \end{pmatrix}$$

$$\dfrac{dT(t)}{dt} = \begin{pmatrix} \dfrac{d^2x(t)}{dt^2} + a\dfrac{dx(t)}{dt} \\ b\dfrac{d^3x(t)}{dt^3} \\ 2cx(t)\dfrac{dx(t)}{dt} \end{pmatrix}$$

- **Interaction Coupling Strength Matrix:**

$$C(t) = \begin{pmatrix} 0 & \dfrac{dx(t)}{dt} & ax(t) \\ 0 & 0 & b\dfrac{d^2x(t)}{dt^2} \\ 0 & 0 & cx^2(t) \end{pmatrix}$$

$$\dfrac{dC(t)}{dt} = \begin{pmatrix} 0 & \dfrac{d^2x(t)}{dt^2} & a\dfrac{dx(t)}{dt} \\ 0 & 0 & b\dfrac{d^3x(t)}{dt^3} \\ 0 & 0 & 2cx(t)\dfrac{dx(t)}{dt} \end{pmatrix}$$

Question 10. If in the previous question the original input source $x(t)$ had the following relation with time:

$$x(t) = x_0 + e^{-2t}$$

Find out the generator Trigger Parameters, Interaction Parameter Coupling Strength Matrix, and their changes through time.

Answer 10:
We get,

- **Trigger Parameter Vector:**

$$T(t) = \begin{pmatrix} T_1 = -2e^{-2t} + a(x_0 + e^{-2t}) \\ T_2 = 4be^{-2t} \\ T_3 = c(x_0 + e^{-2t})^2 \end{pmatrix}$$

$$\frac{dT(t)}{dt} = e^{-2t} \begin{pmatrix} (4 - 2a) \\ -8b \\ -4c(x_0 + e^{-2t}) \end{pmatrix}$$

- **Interaction Coupling Strength Matrix:**

$$C(t) = \begin{pmatrix} 0 & -2e^{-2t} & a(x_0 + e^{-2t}) \\ 0 & 0 & 4be^{-2t} \\ 0 & 0 & c(x_0 + e^{-2t})^2 \end{pmatrix}$$

$$\frac{dC(t)}{dt} = \begin{pmatrix} 0 & 4e^{-2t} & -2ae^{-2t} \\ 0 & 0 & -8be^{-2t} \\ 0 & 0 & -4c\,e^{-2t}(x_0 + e^{-2t}) \end{pmatrix}$$

Or:

$$\frac{dC(t)}{dt} = e^{-2t} \begin{pmatrix} 0 & 4 & -2a \\ 0 & 0 & -8b \\ 0 & 0 & -4c(x_0 + e^{-2t}) \end{pmatrix}$$

Question 11. In the previous question, find out the Trigger Parameters, Interaction Parameter Strength Matrix, and their changes through time as t becomes very large or goes to infinity.

Answer 11:
We get,

- **Trigger Parameter Vector:**

$$\lim_{t \to \infty} T(t) = \begin{pmatrix} ax_0 \\ 0 \\ cx_0^2 \end{pmatrix}$$

$$\lim_{t \to \infty} \frac{d\mathbf{T}(t)}{dt} = \begin{pmatrix} 0 \\ 0 \\ 0 \end{pmatrix}$$

- **Interaction Coupling Strength Matrix:**

$$\lim_{t \to \infty} \mathbf{C}(t) = \begin{pmatrix} 0 & 0 & ax_0 \\ 0 & 0 & 0 \\ 0 & 0 & cx_0^2 \end{pmatrix}$$

$$\lim_{t \to \infty} \frac{d\mathbf{C}(t)}{dt} = \begin{pmatrix} 0 & 0 & 0 \\ 0 & 0 & 0 \\ 0 & 0 & 0 \end{pmatrix}$$

Question 12. Now, if t goes to zero for the previous question, find out the agent Trigger Parameters, Interaction Coupling Strength Matrix, and their changes over time.

Answer 12:
We get,

- **Trigger Parameter Vector:**

$$\lim_{t \to 0} \mathbf{T}(t) = \begin{pmatrix} T_1 = -2 + a(x_0 + 1) \\ T_2 = 4b \\ T_3 = c(x_0 + 1)^2 \end{pmatrix}$$

$$\lim_{t \to 0} \frac{d\mathbf{T}(t)}{dt} = \begin{pmatrix} (4 - 2a) \\ -8b \\ -4c(x_0 + 1) \end{pmatrix}$$

- **Interaction Coupling Strength Matrix:**

$$\lim_{t \to 0} \mathbf{C}(t) = \begin{pmatrix} 0 & -2 & a(x_0 + 1) \\ 0 & 0 & 4b \\ 0 & 0 & c(x_0 + 1)^2 \end{pmatrix}$$

$$\lim_{t \to 0} \frac{d\mathbf{C}(t)}{dt} = \begin{pmatrix} 0 & 4 & -2a \\ 0 & 0 & -8b \\ 0 & 0 & -4c(x_0 + 1) \end{pmatrix}$$

Question 13. If a, b, and c are all equal to 1, for the previous question (for t going to zero), find out the agent Trigger Parameters, Interaction Coupling Strength Matrix, and their changes through time.

Answer 13:
We get,

- **Trigger Parameter Vector:**

$$\lim_{t \to 0} T(t) = \begin{pmatrix} T_1 = (x_0 - 1) \\ T_2 = 4 \\ T_3 = (x_0 + 1)^2 \end{pmatrix}$$

$$\lim_{t \to 0} \frac{dT(t)}{dt} = \begin{pmatrix} 2 \\ -8 \\ -4(x_0 + 1) \end{pmatrix}$$

- **Interaction Coupling Strength Matrix:**

$$\lim_{t \to 0} C(t) = \begin{pmatrix} 0 & -2 & (x_0 + 1) \\ 0 & 0 & 4 \\ 0 & 0 & (x_0 + 1)^2 \end{pmatrix}$$

$$\lim_{t \to 0} \frac{dC(t)}{dt} = \begin{pmatrix} 0 & 4 & -2 \\ 0 & 0 & -8 \\ 0 & 0 & -4(x_0 + 1) \end{pmatrix}$$

Question 14. In question 9, as the system dynamic (SD) equation showed, we have:

$$\frac{d^2x(t)}{dt^2} - \left(\frac{c}{b}\right)x^2(t) + \left(\frac{a}{b}\right)x(t) = 0$$

It is a 2nd order, 2nd degree, Continuous, Non-linear, Non-autonomous, deterministic equation of a CAS system. Explain the general nature of this dynamic equation concerning constant parameters a, b, and c.

Answer 14:
The value and relationship between the three constants, a, b, and c, can impact the dynamic behavior of this CAS. Certain combinations could cause the dynamics to become chaotic, and others make it very stable. Hence, the three constants' analysis can be an essential part of

understanding the resiliency, robustness, and stability of the CAS. Analysis of nonlinear dynamics and differential equations deals with such topics. The concept of chaotic dynamics and stability of a system is also related to such a topic.

Question 15. Assume that the relationship between constants a, b, and c in our CAS in question 9, are such that the SD equation has the following output observations of R_i and σ_i in various output samples i. The following table shows all the observations made. Calculate the Hurst exponents? What is the final state of the system?

Calculation of the Hurst Exponent for the output of the system in question 9						
R_i	σ_i	$\frac{R_i}{\sigma_i}$	$\ln(\frac{R_i}{\sigma_i})$	i	$\ln i$	H
2.2	1.80			10		
2.5	1.90			20		
3	2.20			30		
5	3.00			40		
7	3.60			50		
9	4.50			60		
11	5.50			70		
12	6.00			80		
14	6.70			90		
16	7.20			100		

Answer 15:
The following table shows the results.

Calculation of the Hurst Exponent for the output of the system in question 9						
R_i	σ_i	$\frac{R_i}{\sigma_i}$	$\ln(\frac{R_i}{\sigma_i})$	i	$\ln i$	H
2.2	1.80	1.22	0.20	10	2.30	0.09
2.5	1.90	1.32	0.27	20	3.00	0.09
3	2.20	1.36	0.31	30	3.40	0.09
5	3.00	1.67	0.51	40	3.69	0.14
7	3.60	1.94	0.66	50	3.91	0.17
9	4.50	2.00	0.69	60	4.09	0.17
11	5.50	2.00	0.69	70	4.25	0.16
12	6.00	2.00	0.69	80	4.38	0.16
14	6.70	2.09	0.74	90	4.50	0.16
16	7.20	2.22	0.80	100	4.61	0.17

We know that If $0 \leq H < 0.5$, the CAS output has a tendency to flip flop or fluctuate, or if the system output is increasing, it is more likely to decrease, and if it is decreasing, it is more likely to increase next. In our CAS, as the sample sizes grow from 10 to 100, the Hurst coefficient stays in a range of 0.09 to 0.17. Therefore we have a stable output behavior with a bit of "anti-persistence random walk" or a "Pink noise" dynamics.

Question 16. Now, if the relationship between constants *a, b,* and *c* in our CAS in question 9, are such that the SD equation has output observation results of δ_i and δ_{i-1} as shown in the following table, calculate and interpret the total Lyapunov exponent?

Calculation of the Lyapunov Exponent for the output of a system in question 9					
$\delta(t_i)$	$\delta(t_{i-1})$	$\ln(\frac{\delta(t_t)}{\delta(t_{t-1})})$	t	$\frac{1}{t}$	λ_t
0.004	0.005		10		
0.005	0.007		10		
0.007	0.010		10		
0.010	0.015		10		
0.015	0.023		10		
0.023	0.035		10		
0.035	0.054		10		
0.054	0.085		10		
0.085	0.135		10		
0.135	0.253		10		

Answer 16:
The following table shows the results:

Calculation of the Lyapunov Exponent for the output of a system in question 9					
$\delta(t_i)$	$\delta(t_{i-1})$	$\ln(\frac{\delta(t_t)}{\delta(t_{t-1})})$	t	$\frac{1}{t}$	λ_t
0.004	0.005	-0.223	10	0.100	-0.022
0.005	0.007	-0.336	10	0.100	-0.034
0.007	0.010	-0.357	10	0.100	-0.036
0.010	0.015	-0.405	10	0.100	-0.041
0.015	0.023	-0.427	10	0.100	-0.043
0.023	0.035	-0.420	10	0.100	-0.042
0.035	0.054	-0.434	10	0.100	-0.043

0.054	0.085	-0.454	10	0.100	-0.045
0.085	0.135	-0.463	10	0.100	-0.046
0.135	0.253	-0.628	10	0.100	-0.063
Sum		-4.147			
λ		-0.415			

This CAS has a total negative Lyapunov exponent equal to -0.415. The Lyapunov exponent fluctuates and shows a general tendency of the system output movement towards stability and non-chaotic dynamics through time. As we know, **If $\lambda < 0$**, the average rate of infinitesimal change between two adjacent system outputs is converging to zero. We observe this type of system output dynamics when internal damping, negative feedback, and feedforward structures lead to stable or attractor output results over time.

Question 17. Utilizing the measurement parameter table, analyze the degree of CAS for the movement patterns of 10,000 Bees to protect and maintain their beehive. While functioning, each Bee interacts dynamically with eight other Bees in its immediate vicinity. Provide recommendations in terms of how to enhance the CAS dynamics.

Answer 17:
The results are shown in the following table:

Complex Adaptive System
10,000 Bees in their beehive
"Relative Measurement Parameters"

Parameter	Code	Wt.	Measurement Metrics for 10,000 Bees in their beehive
CAS Type	CT	-	Blended (Physical, Logical)
Number of Agents	NOA	1	Very High (10)
Degree of Non-homogeneity of Agents	DNHA	3	Low (3), Three types of bees
Average Number of Connections per Agents	ANCA	2	High (8)
Average Number of In Connections per Agent	ANICA	1	High (8)
Average Number of Out Connections per Agent	ANOCA	1	High (8)
Agent Coupling Strengths	ACS	2	Medium (4)
Number of Hamiltonian Cycles	NHC	3	Medium (5)

Average System Information Entropy	ASIP	-1	Medium (5)
Average System Information Energy	ASIG	2	Medium (5)
System Dynamics	SD	2	Complex (6)
Degree of Contagion Effect	DCE	3	Very High (8)
Degree of Collective Intelligence	DCI	3	Medium (5)
Degree of Emergence & Innovation	DEI	3	Medium (5)
Degree of Resilience	DRE	1	High (6)
Degree of Robustness	DRO	1	High (6)
Weighted Average Degree of CAS	WADC		Medium (5.8)

The WADC for 10,000 Bees to protect and maintain their beehive is 5.8 points or Medium. To increase the degree of CAS for the Bees, the following actions could be taken:

✓ Create an open beehive where bees can be in more contact to increase the ACS, NHC, and DCE, causing more DCI, DEI.

Question 18. We have a closed information processing system that, after several steps, produces a final stream of data. It has a continuous input data stream which can vary with time represented by $x(t)$. We have three processing units (agents or nodes), each making certain transformations on their data inputs, as shown in the system graph below. The first agent creates two outputs; one equals the first derivative of the initial input with regards to time or $\frac{dx(t)}{dt}$ times constant a and the second as only the first derivative of the initial input with regards to time or $\frac{dx(t)}{dt}$. The second node takes the $\frac{dx(t)}{dt}$ input and generates an output as a second derivative with regards to time or $\frac{d^2x(t)}{dt^2}$ times a constant b. The constants a and b are system factors that can be determined by design. Let us assume those interactions between the first and the second nodes when added linearly, create a final output equal to $x^3(t)$ times a constant c. For various values of a, b, and c, we will get different system dynamic behaviors. The number of processing units and their links are all fixed and do not change. Determine the system's internal network classification, attributes, and interaction coupling strength matrix.

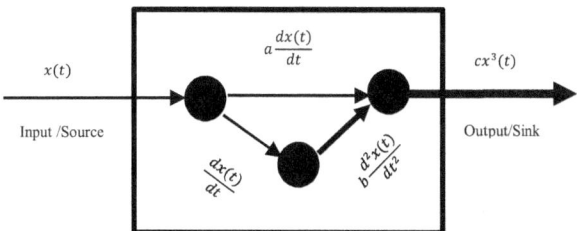

Closed System

Answer 18:

The agent-agent network classification, attributes and the interaction coupling strength matrix we have:

- **System Graph:** Shown above
- **CAS Type:** Information
- **System Attributes:**
  - ✓ NOA — 3 Agents
  - ✓ DNHA — Non-homogeneous
  - ✓ ANCA — 5 links with an avg. 1.7 per agent including the source and sink
  - ✓ ANICA — 4 links with an avg. 1.3 per agent
  - ✓ ANOCA — 4 links with an avg. 1.3 per agent
  - ✓ ACS — Shown on the graph
  - ✓ NHC — 0
  - ✓ ASIP — 0 with no variations in node behaviors
  - ✓ SD — $b\frac{d^2x(t)}{dt^2} + a\frac{dx(t)}{dt} - cx^3(t) = 0$ this is a 2nd order, 3rd degree, Continuous, Non-linear, Non-autonomous, deterministic equation of the system. Parameters a, b, and c are constants.

- **Input/Source:** $x(t)$
- **Output/Sink Result:** Agent3 output $cx^3(t)$
- **Trigger Parameter Vector:**

$$T(t) = \begin{pmatrix} T_1 = \frac{dx(t)}{dt}(a+1) \\ T_2 = b\frac{d^2x(t)}{dt^2} \\ T_3 = b\frac{d^2x(t)}{dt^2} + a\frac{dx(t)}{dt} = cx^3(t) \end{pmatrix}$$

$$\frac{d\boldsymbol{T}(t)}{dt} = \begin{pmatrix} \frac{d^2x(t)}{dt^2}(a+1) \\ b\frac{d^3x(t)}{dt^2} \\ 3cx^2(t)\frac{dx(t)}{dt} \end{pmatrix}$$

- **Interaction Coupling Strength Matrix:**

$$\boldsymbol{C}(t) = \begin{pmatrix} 0 & \frac{dx(t)}{dt} & a\frac{dx(t)}{dt} \\ 0 & 0 & b\frac{d^2x(t)}{dt^2} \\ 0 & 0 & cx^3(t) \end{pmatrix}$$

$$\frac{d\boldsymbol{C}(t)}{dt} = \begin{pmatrix} 0 & \frac{d^2x(t)}{dt^2} & a\frac{d^2x(t)}{dt^2} \\ 0 & 0 & b\frac{d^3x(t)}{dt^2} \\ 0 & 0 & 3cx^2(t)\frac{dx(t)}{dt} \end{pmatrix}$$

Question 19. If in our information processing system in question 19, the input data source $x(t)$ had the following relation with time:

$$x(t) = x_0 e^{-t}$$

Find out the Trigger Parameters, Interaction Parameter Coupling Strength Matrix, and their changes through time.

Answer 19:
We get,

- **Trigger Parameter Vector:**

$$\boldsymbol{T}(t) = \begin{pmatrix} T_1 = \frac{dx(t)}{dt}(a+1) = -x_0 e^{-t}(a+1) \\ T_2 = b\frac{d^2x(t)}{dt^2} = bx_0 e^{-t} \\ T_3 = cx_0^3 e^{-3t} \end{pmatrix}$$

$$\frac{dT(t)}{dt} = \begin{pmatrix} x_0 e^{-t}(a+1) \\ -bx_0 e^{-t} \\ -3cx_0^3 e^{-3t} \end{pmatrix}$$

- **Interaction Coupling Strength Matrix:**

$$C(t) = \begin{pmatrix} 0 & -x_0 e^{-t} & -ax_0 e^{-t} \\ 0 & 0 & bx_0 e^{-t} \\ 0 & 0 & cx_0^3 e^{-3t} \end{pmatrix}$$

$$\frac{dC(t)}{dt} = \begin{pmatrix} 0 & x_0 e^{-t} & ax_0 e^{-t} \\ 0 & 0 & -bx_0 e^{-t} \\ 0 & 0 & -3cx_0^3 e^{-3t} \end{pmatrix}$$

Question 20. In the previous question, find out the Trigger Parameters, Interaction Parameter Strength Matrix, and their changes through time as t becomes very large or goes to infinity.

Answer 20:
We get,

- **Trigger Parameter Vector:**

$$\lim_{t \to \infty} T(t) = \begin{pmatrix} 0 \\ 0 \\ 0 \end{pmatrix}$$

$$\lim_{t \to \infty} \frac{dT(t)}{dt} = \begin{pmatrix} 0 \\ 0 \\ 0 \end{pmatrix}$$

- **Interaction Coupling Strength Matrix:**

$$\lim_{t \to \infty} C(t) = \begin{pmatrix} 0 & 0 & 0 \\ 0 & 0 & 0 \\ 0 & 0 & 0 \end{pmatrix}$$

$$\lim_{t \to \infty} \frac{dC(t)}{dt} = \begin{pmatrix} 0 & 0 & 0 \\ 0 & 0 & 0 \\ 0 & 0 & 0 \end{pmatrix}$$

Question 21. If t goes to zero for question 20, find out the agent Trigger Parameters, Interaction Coupling Strength Matrix, and their changes through time.

Answer 21:
We get,

- **Trigger Parameter Vector:**

$$\lim_{t \to 0} T(t) = \begin{pmatrix} -x_0(a+1) \\ bx_0 \\ cx_0^3 \end{pmatrix}$$

$$\lim_{t \to 0} \frac{dT(t)}{dt} = \begin{pmatrix} x_0(a+1) \\ -bx_0 \\ -3cx_0^3 \end{pmatrix}$$

- **Interaction Coupling Strength Matrix:**

$$\lim_{t \to 0} C(t) = \begin{pmatrix} 0 & -x_0 & -ax_0 \\ 0 & 0 & bx_0 \\ 0 & 0 & cx_0^3 \end{pmatrix}$$

$$\lim_{t \to 0} \frac{dC(t)}{dt} = \begin{pmatrix} 0 & x_0 & ax_0 \\ 0 & 0 & -bx_0 \\ 0 & 0 & -3cx_0^3 \end{pmatrix}$$

Question 22. In our information processing system in question 19, the relationship between constants a, b, and c are such that the SD equation has the output observations of R_i and σ_i in various output samples i as shown in the following table. Calculate the Hurst exponents? What is the final state of this CAS?

Calculation of the Hurst Exponent for the output of the system in question 19							
R_i	σ_i	$\frac{R_i}{\sigma_i}$	$\ln(\frac{R_i}{\sigma_i})$	i	$\ln i$	H
70	3.50			50		
90	3.30			75		
120	3.50			100		
150	3.80			125		

175	4.00			150		
200	4.50			175		
225	5.50			200		
250	6.00			225		

Answer 22:
The following table shows the results.

Calculation of the Hurst Exponent for the output of the system in question 19							
R_i	σ_i	$\dfrac{R_i}{\sigma_i}$	$\ln(\dfrac{R_i}{\sigma_i})$	i	$\ln i$	H	
70	3.50	20.00	3.00	50	3.91	0.77	
90	3.30	27.27	3.31	75	4.32	0.77	
120	3.50	34.29	3.53	100	4.61	0.77	
150	3.80	39.47	3.68	125	4.83	0.76	
175	4.00	43.75	3.78	150	5.01	0.75	
200	4.50	44.44	3.79	175	5.16	0.73	
225	5.50	40.91	3.71	200	5.30	0.70	
250	6.00	41.67	3.73	225	5.42	0.69	

Our CAS output's Hurst exponent value has stayed between 0.77 and 0.69, which indicates a *"persistence random walk"* or a *"Black noise"* behavior. If our system output is increasing, it is more likely to increase next, and if decreasing, it is more likely to decrease next. This behavior potentially tends to lead to chaotic behavior of the output. Our information processing system has more correlation or dependencies between the system outputs over time. The closer the value of H gets to 1, the more possibility exists to have output instabilities and observing chaotic system behavior. Our H value is not close to 1, and it is reducing at larger sample sizes, indicating a more stable output result for larger samples.

Question 23. If the relationship between constants a, b, and c in our CAS in question 19, are such that the SD equation has output observation results of δ_i and δ_{i-1} as shown in the following table, calculate the total Lyapunov exponent? How do we interpret the behavior of the system output?

Calculation of the Lyapunov Exponent for the output of a system in question 19						
$\delta(t_i)$	$\delta(t_{i-1})$	$ln(\frac{\delta(t_t)}{\delta(t_{t-1})})$	t	$\frac{1}{t}$	λ_t
9.000	7.550		10		
9.100	7.600		10		
9.300	7.800		10		
9.450	7.900		10		
9.550	7.950		10		
9.650	8.000		10		
9.700	8.050		10		
9.750	8.100		10		
9.800	8.100		10		
9.850	8.050		10		

Answer 23:

The following table shows the results:

Calculation of the Lyapunov Exponent for the output of a system in question 19					
$\delta(t_i)$	$\delta(t_{i-1})$	$ln(\frac{\delta(t_t)}{\delta(t_{t-1})})$	t	$\frac{1}{t}$	λ_t
9.000	7.550	0.176	10	0.100	0.018
9.100	7.600	0.180	10	0.100	0.018
9.300	7.800	0.176	10	0.100	0.018
9.450	7.900	0.179	10	0.100	0.018
9.550	7.950	0.183	10	0.100	0.018
9.650	8.000	0.188	10	0.100	0.019
9.700	8.050	0.186	10	0.100	0.019
9.750	8.100	0.185	10	0.100	0.019
9.800	8.100	0.191	10	0.100	0.019
9.850	8.050	0.202	10	0.100	0.020
Sum		1.846			
λ		0.185			

The total Lyapunov exponent of this CAS is equal to 0.185. The Lyapunov exponent λ is positive and δ_t is growing, and the systems' output is moving towards becoming unstable. This type of output dynamics is usually observed when a system leads to volatile, *"amplifying,"* unstable, and *"chaotic"* output results over time.

Question 24. It is known that Sardines group together when they are threatened. Utilizing the measurement parameter table, analyze the degree of CAS for the Swarm movement patterns of 1,000,000 Sardines in their yearly migration in the South African waters. While functioning, assume each sardine interacts dynamically with eight others, and there are 18 different types of Sardines in the group.

Answer 24:

The results are shown in the following table:

Complex Adaptive System
Swarm of 1,000,000 Sardines in the Ocean
"Relative Measurement Parameters"

Parameter	Code	Wt.	Measurement Metrics for Swarm of 1,000,000 Sardines
CAS Type	CT	-	Blended (Physical, Logical)
Number of Agents	NOA	1	Very High (10)
Degree of Non-homogeneity of Agents	DNHA	3	Very High (10), 18 types of Sardines
Average Number of Connections per Agents	ANCA	2	High (8)
Average Number of In Connections per Agent	ANICA	1	High (8)
Average Number of Out Connections per Agent	ANOCA	1	High (8)
Agent Coupling Strengths	ACS	2	High (8)
Number of Hamiltonian Cycles	NHC	3	Medium (5)
Average System Information Entropy	ASIP	-1	Medium (5)
Average System Information Energy	ASIG	2	Medium (5)
System Dynamics	SD	2	Complex (6)
Degree of Contagion Effect	DCE	3	Medium (5)
Degree of Collective Intelligence	DCI	3	Medium (5)
Degree of Emergence & Innovation	DEI	3	Medium (5)
Degree of Resilience	DRE	1	Medium (5)
Degree of Robustness	DRO	1	Medium (5)
Weighted Average Degree of CAS	WADC		High (6.5)

It is a SIS with a high WADC equal to 6.5.

Question 25. To resolve a regional conflict such as the one in the Syrian crisis in the Middle East, provide some suggestions regarding where the Swarm Intelligence Structure could be utilized?

Answer 25:
Swarm Intelligence applications can help a group of agents (state leaders, officials, or countries) prevent a shock, defend against an attack, find food, or an optimum solution to a common problem. The regional conflict in the Middle East needs to be resolved in a mutually acceptable manner. The Swarm Intelligence approach can be used in international committees with representatives from each country to determine a mutually accepted conflict resolution. In conflict resolution committees utilizing Swarm Intelligence Structures, the challenge would be determining proper agent-agent rules and algorithms to minimize potential errors and risks to attain accepted intended goals and solutions. The complexity of individual egos, negative memories, sentiments and perceptions, and domestic politics of each side create frictions and make the process more complicated.

Question 26. Which of the following behaviors utilize Swarm Intelligence Structure?
- A large number of Locust searching for food.
- A Shark is attacking a seal.
- A Whale is moving in the ocean.

Answer 26:
The dynamics of Swarm Intelligence is based on AIAR behavior which is summarized as follows:
- Each cognitive agent communicates, interacts, and is aware of its surroundings – **Awareness**
- Each cognitive agent is autonomous and makes independent decisions – **Independence**
- Each cognitive agent is aware of what it can do – **Autonomy**
- Cognitive agents can be added, eliminated, and replaced with no effect on the system – **Resilience**

The above features can be observed when a large number of Locust are searching for food

Question 27. As the agent's cognitive abilities in a Swarm Intelligence Structure increase, how do your CAS features change?

Answer 27:
As the agent cognitive abilities in a Swarm Intelligence Structure increases, ACS, ASIG, SD, DCE, DCI, DEI, DRE, and DRO increase. These increases lead to a higher WADC. If ACS reaches a critical threshold value for behaving synchronized, we will also observe Synchronized Swarm behavior.

Question 28. As the degree of agent-agent interactions or ANCA increases in a Swarm Intelligence Structure, how do the following CAS features change?
- System Dynamics, SD
- Degree of contagion effects, DCE
- Weighted Average Degree of CAS

Answer 28:
- System Dynamics will become more complex
- The degree of contagion effects will increase
- WADC will increase

Question 29. What can increase the Average System Information Entropy in a SIS?

Answer 29:
For the total system, the Average System Information Entropy is defined as;

$$ASIP = -(\sum_{i=1}^{i=n} \sum_{j=1}^{j=j} p_j^i \ln p_j^i)/n$$

For each agent in the system, the summation of probabilities of observing each state p_j^i must add up to 1 or:

$$\sum_{j=1}^{j=j} p_j^i = 1$$

As the individual observed probability for an agent or p_j^i and collectively, for all agents to become larger and closer to 1, *ASIP* gets closer to 0. With more certainty, less randomness, or more structure, we have less information entropy in the system.

Whatever increases the SIS's uncertainty and degree of behavioral randomness will increase Average System Information Entropy or ASIP. Any reductions in the agent's cognitive abilities, coupling, independence, autonomy, system resilience, and robustness will cause reduced structures, fewer algorithms, and limitations on its behavior. It will also increase the ASIP and reduce ASIG of a Swarm Intelligence Structure.

Question 30. We have two CASs. Both have highly cognitive agents (the same *Aep*) and the same NOA and DNHA. The first one has high ANCA (resulting in high *Aea*) and the second one has low ANCA (resulting in low *Aea*). Show how this affects the relative values of ASIG and ASIP.

Answer 30:
The results are summarized in the following table:

CAS	ASIG	ASIP
First	Larger	Smaller
Second	Smaller	Larger

Due to larger ASIG and smaller ASIP, the first CAS will probably be more adaptive, retain memory, intelligent, innovative, emergent, resilient, and robust.

Question 31. We have two CASs. Both have the same ANCA (the same *Aea*), NOA, and DNHA. The first one has a high cognitive ability (resulting in high *Aep*), and the second one has a low cognitive ability (resulting in low *Aep*). Show how this affects the relative values of ASIG and ASIP.

Answer 31:
The results are summarized in the following table:

CAS	ASIG	ASIP
First	Larger	Smaller
Second	Smaller	Larger

As in question 31, and larger ASIG and smaller ASIP, the first CAS will probably be more adaptive, retain memory, intelligent, innovative, emergent, resilient, and robust.

Question 32. How do the information structural symmetries of the two CASs in questions 31 and 32 differ?

Answer 32:
In both questions, the ASIG is smaller and ASIP larger for the second CASs. The second CAS in both questions has a smaller information structure, larger information entropy, and larger information structural symmetries.

Question 33. We have two brilliant brokerage organizations. Both have the same diversity and highly cognitive traders (the same DNHA, NOA, and *Aep*). The first one has sophisticated information and an interactive network of traders (higher ANCA or *Aea*). The second one has a more traditional paper-based information and interactive network of traders (lower ANCA or *Aea*). Show how this affects the relative values of ASIG and ASIP.

Answer 33:
The results are summarized in the following table:

CAS	ASIG	ASIP
First Brokerage	Larger	Smaller
Second Brokerage	Smaller	Larger

Due to larger ASIG and smaller ASIP, the first Brokerage firm will show more adaptive, learning, memory, intelligent, innovative, emergent, resilient, and robust behavior in its trading functions.

Question 34. How do the information structural symmetries of the two Brokerage houses in the previous question differ?

Answer 34:
The second Brokerage house has more information structural symmetries with a smaller ASIG or average system information energy and larger ASIP or average system information entropy.

Question 35. Briefly explain the relationship between the complex structure of a CAS incorporating high levels of feedback, feedforward, NHC, DCE, DCI, DEI, and the structural symmetry and ASIG of a system.

Answer 35:
Higher levels of feedback, feedforward, NHC, DCE, DCI, and DEI in a CAS means that there exists a very complex agent-agent relationship and ingrained structure or high ASIG in place. At the same time, the structural symmetry and ASIP will be lower.

Complex Adaptive Financial Systems

Definition

Financial System is a platform for the exchange, intermediation, and balance of the sources and uses of funds. It allows funds transfer from individuals or institutions with excess and available funds to individuals or institutions needing funds. This system is composed of a set of interrelated and interconnected Institutions, Instruments, Markets, and Rules & Regulations.

- **Financial Institutions (FI)**

    They are regulated legal entities[48], each with specific functions and roles as financial intermediations in the exchange and balance of funds. The main types of institutions include:

    - ***Banks or Deposit Based Institutions***

        They fulfill the financial intermediation role by accepting funds in various deposits (from those with available funds)

---

[48] From a CAS perspective, these legal entities are operated by responsible and cognitive individuals or agents following certain strict internal policies & procedures and external regulatory frameworks which define and shape the ANCA, ANICA, ANOCA and ACS (sometimes sub-optimal) parameters. Financial Institutions are specialized clusters of cognitive individuals with specific cluster tasks and goals and high degrees of DNHA, ANCA, ANICA, ANOCA, ACS, NHC, ASIG, DCE, low ASIP and complex SD. From a regulators point of view the hope is that these clusters would have high DCI, DEI, DRE, and DRO. In the past and at certain times (financial system collapses) the sub-optimal internal & external rules & regulations and inefficient ANCA, ANICA, ANOCA, ACS, NHC, ASIP, ASIG lead to sub-optimal DCE and inflexible SD causing a low degree of DCI, DEI, DRE and DRO causing the collapse and failure of the system.

and allocating them to those in need through various loans. Major types of banks include:
- ✓ Commercial/Retail Banks
- ✓ Corporate Banks
- ✓ Private Banks
- ✓ Saving & Loan Associations
- ✓ Developmental Banks
- ✓ Export/Import Banks

- *Insurance Institutions*
  They act as intermediaries to cover and manage financial risks and uncertainties. They collect funds in the form of annual premiums and, by pooling and managing them cover future possible financial losses of the covered individuals or institutions. The main types of Insurance institutions include:
  - ✓ Life/Health Insurance
  - ✓ Auto/Mortgage Insurance
  - ✓ Property/Home Insurance
  - ✓ Pension Funds (income insurance for retirement)
  - ✓ Marine Insurance

- *Investment Banks*
  They act as intermediary financial advisors by finding non-deposit-based funds and allocating them to those of need in various forms of equity, debt, or other hybrid types of structures. These particular banks usually incorporate the following functions:
  - ✓ Corporate Finance Advisory & Financial Consulting
  - ✓ Trading & Sales of Securities
  - ✓ Financial Research and Analysis

- *Brokerage Companies*
  They are financial intermediaries involved in brokering the trade, sale, and purchase of financial instruments, primarily for other individuals, institutions, and at times for themselves. In line with this role, they can also act as financial advisors. There are various types of brokerage companies, and the main ones are:
  - ✓ Securities Brokers

- ✓ Commodities Brokers
- ✓ Loan Brokers, including Mortgage Brokers

- *Investment Companies*
  They are financial intermediaries principally involved in managing individual or institutional funds for investments into securities. They can be considered as a form of financial instrument management company. There are three main types of investment companies with very diverse goals and objectives. They include:
  - ✓ Closed-End Investment Management Companies known as closed-end funds
  - ✓ Open-End Investment Management Companies known as open-end or Mutual Funds
  - ✓ Unit Investment Trusts, known as UITs

- *Financial Cooperative Institutions*
  These are financial institutions owned and managed by specific members with common interests. They provide services to their members, similar to other financial institutions, in a more customized and efficient manner. They have less emphasis on profitability, focusing on efficient and customized services to their members/owners. The main types are:
  - ✓ Credit Cooperatives, known as Credit Unions
  - ✓ Insurance Cooperatives
  - ✓ Investment Cooperatives

- **Financial Instruments/Securities (FS)**
  These are legal and regulated financial contracts that transfer funds, wealth, or asset ownership in the markets by individuals or institutions. Financial instruments carry Security types are very diverse, but the main ones are:
  - ✓ Equity-Based Securities (e.g., Common, Preferred Stocks)
  - ✓ Debt-Based Securities (e.g., Bills, Notes, Bonds, Debentures)
  - ✓ Derivative Based Securities (e.g., Forwards, Futures, Options, Swaps)
  - ✓ Hybrid Based Securities (e.g., Swaptions, Optures)
  - ✓ Commodities Based Securities

- ✓ Currencies (we will assume that this is security)

- **Financial Markets (FM)**
  They are legal and regulated platforms (physical or virtual) to trade diverse financial instruments. Markets are where the transfer of funds occurs. The players are the individuals or institutions, and the means are the financial instruments or securities. There are various types of markets, and they include:
    - ✓ Based on instrument differentiations
        - Equity Markets
        - Debt Markets
        - Derivative Markets
        - Commodities Markets
        - Currency Markets
    - ✓ Based on instrument age
        - Initial Public Offerings or IPO market
        - Secondary Market
    - ✓ Based on regulatory structures of the markets
        - Highly regulated or Exchanges
        - Less regulated or Over the Counter (OTC)

- **Financial Rules & Regulations (FRR)**
  Specific rules and regulations are required to create fair, orderly, and equitable markets for trading instruments among individuals or institutions. Rules & regulations organize and focus on the institutional, markets, and securities structures and behaviors. Depending on the types of governments and the existing regulatory structures (centralized versus decentralized), the rules & regulations can vary and are very diverse. In general, the financial rules & regulations should cover:
    - ✓ Financial Institutions
        - Banks
        - Insurance Companies
        - Investment Banks
        - Brokerage Companies
        - Investment Companies
        - Financial Cooperatives
    - ✓ Financial Instruments/Securities
        - Equities Based Securities
        - Debt Based Securities

- - Derivative Based Securities
  - Commodities Based Securities
  - Currencies
  - ✓ Financial Markets
    - Exchanges
    - Over the Counter
    - IPO
    - Secondary Market Trading

- **Financial Regulatory Structure (FRS)**
  Globally, there are two primary financial regulatory structures. Some variants of these two structures are also utilized. For our purpose, we will only deal with the two basic ones. They are as follows:
  - ✓ *Centralized regulatory structure*
    It is based on a single national legal, regulatory entity responsible for coordination, regulation, and monitoring all active financial institutions, markets, and instruments for fair, orderly, and equitable operations. The structure has the benefit of a holistic/helicopter and coordinated regulatory view. The challenge is the difficulties of looking at the regulatory components with focus and specialization.
  - ✓ *Decentralized regulatory structure*
    It is based on several specialized national legal, regulatory entities responsible for regulating and monitoring financial institutions, markets, and instruments separately. This structure's main benefit is more focused and specialized regulatory authorities, and the challenge is the problematic coordinative efforts required for a highly complex and intertwined system.

Financial systems have multiple agents, players, and institutions. This system entails the following dynamics[49]:
- Financial agents are many
- Financial agents are highly cognitive and reactive.
- Financial agents interact continuously, dynamically, and the interactions are financial, informational, and physical.

---

[49] Also see **Shayan, S. A. (2019)**, "Understanding Complex Adaptive Systems," Independent Publisher, Amazon.

- Such interactions are rich, i.e., any agent or sub-agent in the system is affected by and affects several other agents or sub-agents simultaneously.
- The agent interactions are non-linear, meaning small changes in inputs, information, monetary, or physical interactions can create large effects or significant changes in the financial system's behavior as a whole. The self-organized criticality, synchronization, and at times herd phenomenon is observed in financial systems.
- Agent interactions are primarily but not exclusively with immediate neighbors or institutions, and the nature of their influence can get modulated, synchronized, leading to financial bubble creations.
- Agents in the financial system usually behave as social agents. It means that agents (traders, brokers, dealers, investors, institutions, and regulators) can think and initiate change independently. It will be an additional element that can add to the degree of complexity of the systems.
- Financial systems behave in emergent, new, and innovative collective manners.
- Filtering out noise from information as input is a part of coarse-graining, learning, and identifying correct patterns and regularities in the system. It means analysts and specialists can differentiate useful information from noise.
- Any interactions with other agents are feedback onto itself directly or after several intervening stages. Such feedbacks can vary in quality. It is known as the recurrence effect in the financial system, which can cause positive or negative loops in the system's dynamics.
- Developed financial systems do not have any system boundaries. They are open, international, and interact with the global system.
- Developed financial systems show memory, learning, or adaptive properties collectively. They have collective intelligence, making the agents more likely to survive, be resilient, evolve, and become sustainable.
- Financial systems operate in a state of dynamic, continuous, and changing equilibrium.
- Depending on the degree of market efficiency, financial systems have different degrees of history or memory. These systems evolve, and their past is co-responsible for their present and future

behavior. There is a nonlinear correlation between the past, present, and future. It can be interpreted as having memory ingrained in the system's structure, causing ASIP to get lowered and ASIG to get higher.

A fully functional financial system would be a complex interaction of interrelated and interconnected institutions, instruments, and markets through a highly intertwined web of rules and regulations. *Financial systems are certainly complex systems operated by highly cognitive, educated, intelligent, and competitive agents. They are classified as a Complex Adaptive Financial System or CAFS.*

Financial agents (humans) are thoughtful, cognitive, independent, and interact with various degrees in a network of hierarchical connections through various adaptive processes. These connections are usually very complicated, dynamic, and changing, such as those that link bankers, traders, and portfolio managers. CAFS agents think, make independent decisions, predict, react, and adapt to other agents' actions and predictions. They have high degrees of positive and negative nonlinear feedback and feedforward effects on one another in addition to ingrained mutuality structures. The various connections between CAFS agents make them independent and, at the same time, tightly coupled and correlated to one another. It leads to agent-agent interactions becoming highly nonlinear and coevolving, causing changing ANCA, ANICA, ANOCA, NHC, and DCE through positive and negative feedback and feedforward information exchange process, with a high degree of possible contagion behavior leading to possible herd mentality phenomena and very complex and challenging predictability for the CAFS behavior.

Financial theories assume that agents in competitive markets efficiently aggregate information through a price dynamic process leading to single, stable price balancing supplies and demands for the financial instruments and services provided. *Information fluctuations in the markets drive price fluctuations.* In many instances, price fluctuations far outstrip information variations. The existence of adaptive heterogeneous agents, including bankers, insurance experts, traders, portfolio managers, and all buy and sell-side participants, will respond to such fluctuations accordingly. Some CAFS agents will buy, and some will sell, and their collective actions will trigger complex adaptive behavior resulted from feedback, feedforward, and sometimes herd style mentality. If all agents were homogeneous or behaved the same, we could possibly see herd-style dynamics leading to occasional market bubbles and crashes. Heterogeneous cognitive agents with competitive nature would reduce the possibility of correlated herd behavior, and markets would adjust and stay in continuous dynamic equilibrium. It would lead to more adaptive, evolving, and robust dynamics of the markets.[50]

Innovations and Emergence are generally considered processes that lead to the appearance of structure not directly described by the defining constraints and forces that control the system. In CAFS emergence, the localized behavior of agents aggregates into global behavior. There can be multiple layers of financial emergence. One can observe emergence at lower or agent-agent levels or higher agent cluster levels. More heterogeneity and diversity in CAFS agents lead to more flexibility and the emergence of new agent-agent interactions.

Emergence in CAFS is not imposed by a central force or management process but results from an interactive rule-based structure between agents, affecting the system at various levels. Over time, emergent behavior in CAFS leads to innovation and evolution, which can cause new evolving resilient, robust, and adaptive systems.

We will look at some related examples.

Example 1:

> Banks are deposit-taking institutions. They attract small to large funds from individuals and companies in the form of interest-paying deposits. They then pool and manage these deposits for proper allocations to other individuals and companies in need of funds in the form of interest-based loans or mortgages. Through this process, and to protect various stakeholders' interests for fair, orderly, and equitable operations, they must comply with stringent internal policies and procedures and external rules & regulations imposed by the shareholders and the regulators. In the grand scheme of a complex adaptive financial system, we can think of the banking system as a sub-cluster with its structure and dynamics interacting with other sub-clusters of CAFS. We can assume that the banking system in a diversified economy is a complex adaptive open sub-system with individuals (clients and personnel of the banks and regulator) as agents. We will call this sub-system CABS. Assume that we are

---

[50] Ibid.

dealing with a developed country with 20 fully modern, diversified, and highly interconnected banks. There are 20000 employees and 1,000,000 clients as deposit and loan holders, including the regulator's personnel. The banks have the most up-to-date IT and cyber-security systems connected through the internet and regulated in a decentralized manner. It means that there is one central bank in place that acts as the main regulatory body for the 20 banks. Given the description provided and the need to analyze the average five-year loan rate (average for the 20 banks) as the sub-system output, determine the system's internal network classification, attributes, and interaction coupling strength matrix. Next, using the measurement parameter table, determine the degree of complexity of the five-year average loan rate as the CABS' output. Assume that the five-year loan rate of x_i for each bank follows the following equation based on the short term inflation rate of ε, the coefficient for each bank b_i, and x_0 as the base rate:

$$x_i = x_0 b_i (1 + e^{-\varepsilon})^{-1}$$

For the five-year average loan rate or 5YALR, we get:

$$5YALR = (\sum_{i=1}^{i=20} x_0 b_i (1 + e^{-\varepsilon})^{-1})/20$$

Answer 1

The system's graph is shown without coupling interactions, where each bank is shown as ●, and the regulator as 😊 ;

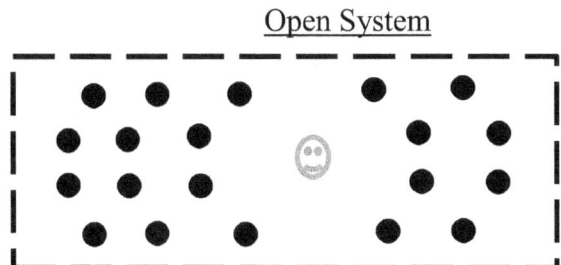

Open System

We have,

- **System Graph:**   Shown above without the links (too many)
- **CAS Type:**   Blended (Physical, Information, Logical)

- **System Attributes:**
  - ✓ NOA   21 institutional agents and 1,020,000 individual agents
  - ✓ DNHA   Non-homogeneous
  - ✓ ANCA   441 institutional links (21 x 21) with an avg. of 21 per institutions and a huge number of links between individual agents (1,020,000 x 1,020,000) with the assumption that an agent can learn or effect itself.
  - ✓ ANICA   441 links with an avg. 21 per agent
  - ✓ ANOCA   441 links with an avg. 21 per agent
  - ✓ ACS   Defined as $c_{ij}(t)$ not shown above
  - ✓ NHC   Many
  - ✓ ASIP, ASIG   Due to the described CABS structure and the internal policies and procedures and the external regulatory limitations, the average system information entropy or ASIP will be low (due to regulatory restrictions) and the average system information energy or ASIG will be high.
  - ✓ SD   Very complex. This is a complex adaptive financial system with 21 institutional agents and trigger parameters that can be represented by a vector with 21 elements of $T_1(t)$ to $T_{21}(t)$ and a 21 x 21 interaction coupling matrix that results in the 5 year average loan rate $5YALR$ as follows (for the sake of simplicity we are ignoring 1,020,000 individual agent interactions and assume that the effects of individual agents for each bank is felt through one interaction from that bank. This means that we are assuming that each bank is a sub-sub-cluster of a bigger banking sub-cluster of CABS, with individual agents as the members):

$$T(t) = \begin{pmatrix} T_1(t) \\ \vdots \\ T_{21}(t) \end{pmatrix}$$

$$C(t) = \begin{pmatrix} c_{11}(t) & \cdots & c_{1\,21}(t) \\ \vdots & \ddots & \vdots \\ c_{21\,1}(t) & \cdots & c_{21\,21}(t) \end{pmatrix}$$

The state of each institutional agent and the interaction coupling strength matrix can mutually and dynamically affect one another and change through time. We have a co-interacting, co-evolving adaptive behavior between agents and their respective interactions. Mathematically this can be expressed in terms of 42 sets of coupled nonlinear partial differential equations. The general mathematical form for this system of equations can be expressed as:

$$\frac{dT_i(t)}{dt} = P\left(c_{ij}(t), T_j(t)\right) \quad for\ i,j\ from\ 1\ to\ 21$$

$$\frac{dc_{ij}(t)}{dt} = Q\left(c_{ij}(t), T_j(t)\right) \quad for\ i,j\ from\ 1\ to\ 21$$

Where P and Q can be different for each bank and are dependent on each institution's strategy[51].

- **Input/Source:** ε (inflation rate)
- **Output/Sink Result:** $(\sum_{i=1}^{i=20} x_0 b_i (1 + e^{-\varepsilon})^{-1})/20$

- **General Interaction Coupling Strength Matrix:**

$$C(t) = \begin{pmatrix} c_{11}(t) & \cdots & c_{1\ 21}(t) \\ \vdots & \ddots & \vdots \\ c_{21\ 1}(t) & \cdots & c_{21\ 21}(t) \end{pmatrix}$$

This matrix changes dynamically through time, and the change can mathematically be represented as:

---

[51] For each bank and the one central bank, we have two equations for a total of the 42 sets of coupled nonlinear partial differential equations as follows:

$$\frac{dT_1(t)}{dt} = P_1\left(c_{1j}(t), T_j(t)\right) \quad for\ j\ from\ 1\ to\ 21\ for\ the\ \mathbf{1st\ agent}$$

$$\frac{dc_{1j}(t)}{dt} = Q_1\left(c_{1j}(t), T_j(t)\right) \quad for\ j\ from\ 1\ to\ 21\ for\ the\ \mathbf{1st\ agent}$$

....

$$\frac{dT_{21}(t)}{dt} = P_{21}\left(c_{21j}(t), T_j(t)\right) \quad for\ j\ from\ 1\ to\ 21\ for\ the\ \mathbf{21st\ agent}$$

$$\frac{dc_{21j}(t)}{dt} = Q_{21}\left(c_{21j}(t), T_j(t)\right) \quad for\ j\ from\ 1\ to\ 21\ for\ the\ \mathbf{21st\ agent}$$

$$\frac{d\boldsymbol{C}(t)}{dt} = \begin{pmatrix} \frac{dc_{11}}{dt} & \cdots & \frac{dc_{1\,21}}{dt} \\ \vdots & \ddots & \vdots \\ \frac{dc_{21\,1}}{dt} & \cdots & \frac{dc_{21\,21}}{dt} \end{pmatrix}$$

The Measurement Parameter is as follows:

Complex Adaptive Banking System
20 developed banks and one central bank
"Measurement Parameters"

Parameter	Code	Wt.	Measurement Metrics
CAS Type	CT	-	Blended (Physical, Information, Logical)
Number of Agents	NOA	1	Very High (10)
Degree of Non-homogeneity of Agents	DNHA	3	Very High (10)
Average Number of Connections per Agents	ANCA	2	Very High (10)
Average Number of In Connections per Agent	ANICA	1	Very High (10)
Average Number of Out Connections per Agent	ANOCA	1	Very High (10)
Agent Coupling Strengths	ACS	2	Very High (10)
Number of Hamiltonian Cycles	NHC	3	Medium (5)
Average System Information Entropy	ASIP	-1	Low (3)
Average System Information Energy	ASIG	2	High (8)
System Dynamics	SD	2	Complex (7)
Degree of Contagion Effect	DCE	3	High (7)
Degree of Collective Intelligence	DCI	3	Medium (5)
Degree of Emergence & Innovation	DEI	3	Low (3)
Degree of Resilience	DRE	1	Medium (5)
Degree of Robustness	DRO	1	Medium (5)
Weighted Average Degree of CAS	WADC		High (7.3)

It is a complex system with a WADC value of 7.3, which is high.

Example 2:

If, in the previous question, the probability distribution of the five-year loan rates x_i for each bank was equal to $P_i(x)$ and was independent of the other banks and followed the following distribution function:

$$P_i(x) = b_i x_0 e^{-0.5\left(\frac{x-\mu_i}{\sigma_i}\right)^2} \quad \text{and} \quad \sum_{x=0}^{x=x} b_i x_0 e^{-0.5\left(\frac{x-\mu_i}{\sigma_i}\right)^2} = 1$$

With an average of μ_i and standard deviation of σ_i for each bank, find out the Average System Information Entropy or ASIP for 5YALR.

Answer 2

- **ASIP equals to:**

$$S_1 = -\sum_{x=0}^{x=x} (b_1 x_0 e^{-0.5\left(\frac{x-\mu_i}{\sigma_i}\right)^2}) \ln (b_1 x_0 e^{-0.5\left(\frac{x-\mu_i}{\sigma_i}\right)^2})$$

$$S_2 = -\sum_{x=0}^{x=x} (b_2 x_0 e^{-0.5\left(\frac{x-\mu_i}{\sigma_i}\right)^2}) \ln (b_2 x_0 e^{-0.5\left(\frac{x-\mu_i}{\sigma_i}\right)^2})$$

....

$$S_{23} = -\sum_{x=0}^{x=x} (b_{20} x_0 e^{-0.5\left(\frac{x-\mu_i}{\sigma_i}\right)^2}) \ln (b_{20} x_0 e^{-0.5\left(\frac{x-\mu_i}{\sigma_i}\right)^2})$$

$$ASIP = \frac{S_1 + S_2 + \cdots + S_{20}}{20}$$

How to manage CAFS

Managing CAFS requires an initial assessment of the system's internal network classification, attributes, and interaction coupling strength matrix. We then use the measurement parameter table to determine the degree of complexity or WADC and the means to manage it.

All CAFS agents are cognitive and can think and learn. They have the autonomy to adapt, coordinate, and synchronize their dynamics with other agents or their environments. *The more autonomous and cognitive CAFS agents are, the more complex and emergent the system behavior will become.*

Based on our Measurement Parameter table, the following table explains various qualitative measures that could be taken to control and manage WADC. The goal is to be able to influence and direct CAFS behavior towards the intended dynamic behavior.

Complex Adaptive Financial System
"How to Manage Complexity through Measurement Parameters"

Parameter	Code	Wt.	How to Increase or Decrease WADC
CAS Type	CT	-	Separate or consolidate the CAFS types.
Number of Agents	NOA	1	Increase or decrease NOA.
Degree of Non-homogeneity of Agents	DNHA	3	Increase or decrease DNHA
Average Number of Connections per Agents	ANCA	2	Increase or decrease ANCA
Average Number of In Connections per Agent	ANICA	1	Increase or decrease ANICA
Average Number of Out Connections per Agent	ANOCA	1	Increase or decrease ANOCA
Agent Coupling Strengths	ACS	2	Increase or decrease ACS by adding or eliminating internal structures or agent-agent feedback/feedforward interactions
Number of Hamiltonian Cycles	NHC	3	Increase or decrease NHC by creating or breaking the NHC loops and internal structures or agent-agent feedback/feedforward interactions
Average System Information Entropy	ASIP	-1	Decrease or increase ASIP by breaking up or adding some internal structures or agent-agent feedback/feedforward interactions
Average System Information Energy	ASIG	2	Increase or decrease ASIG by adding or breaking up some internal structures or agent-agent feedback/feedforward interactions
System Dynamics	SD	2	Increase or decrease the coupling features of the total system of equations. Make the system of equations dependent or independent from one another and less complex.
Degree of Contagion Effect	DCE	3	Increase or decrease DCE by increasing or decreasing ANCA, ANICA, ANOCA, NHC, and adding or breaking up some internal structures or agent-agent feedback/feedforward interactions.
Degree of Collective Intelligence	DCI	3	Increase or decrease DCI by increasing or decreasing ANCA, NANICA, ANOCA, ACS, DCE, adding or breaking up the total number of agents, and maximizing or minimizing cluster-cluster interactions. Also, increase or decrease the number of NHCs and add or break up some internal structures or agent-agent feedback/feedforward interactions.
Degree of Emergence & Innovation	DEI	3	Increase or decrease DEI by increasing or decreasing ANCA, NANICA, ANOCA, ACS, DCE, adding or breaking up the total number of agents, and maximizing or minimizing cluster-cluster interactions. Also, increase or decrease the number of NHCs and add or break up some internal structures or agent-agent feedback/feedforward interactions.
Degree of Resilience	DRE	1	Increase or decrease DRE by increasing or decreasing ANCA, ANICA, ANOCA, ACS, DCE, NHCs and adding or breaking up internal structures or agent-agent feedback/feedforward interactions.

Degree of Robustness	DRO	1	Increase or decrease DRO by increasing or decreasing ANCA, ANICA, ANOCA, ACS, DCE, NHCs and adding or breaking up internal structures or agent-agent feedback/feedforward interactions.
Weighted Average Degree of CAS	WADC		

To achieve the targeted WADC, adjusting the measurement parameters needs experience and a trial and error approach. Due to the highly nonlinear behavior of CAFS, it is impossible to expect a precise cause and effect mechanism in place.

A higher number of cognitive agents, diversity of agents, mutual interaction, connectivity, cooperation, collaboration, and communication (feedback/feedforward loops) are significant contributors to Complex Adaptive Financial behavior. More nonlinearity causes additional complexity, and more learning and cognitive abilities can cause more adaptive dynamics in CAFS. The more robust a CAFS is, the quicker the corrective dynamics towards equilibrium will be reached.

Financial incentives in CAFS cognitive agents trigger faster learning and allow the system to reach complex adaptive levels faster. It should be clear that in less diverse, more centralized, and controlled financial systems, the resulting dynamics are less complex, and CAFS benefits such as robustness, resiliency, adaptability, dynamic equilibrium, emergence, and innovations will be less observed.

With an increase in globalization and global connections, population awareness, free-market emphasis, increased competition, information, noise, communication, internet, the financial system has gotten more complex, unpredictable, with lots of moving parts and volatility of perspectives and opinions of the institutions, markets and instruments involved. This degree of increased complexity is even more prevalent for multi-national financial institutions. These institutions function in many countries and deal with various products, markets, consumers, cultures, regulatory regimes, multi-dimensional uncertainties, and risks.

Higher-risk or information uncertainty in any financial system can be viewed as having higher ASIP. Information uncertainty increases if entropy is not managed, which leads to higher risks and volatilities. One must regulate and manage the information uncertainty of a financial system at the desired levels to contain its risks and entropy. Information uncertainty can exist at the agent-agent up to system levels, collectively.

Managing systematic risks and uncertainties of a financial system is a central part of all regulatory strategies. Based on CAFS management concepts, properly managing a financial system's diversity in terms of the number of agents, nature of feedback, and feedforward relationships among agents reduces the potential uncertainties and risks. It can also trigger innovation, creativity and develop new frontiers through possible synergies and shared values that get created through this process.

Global financial systems are in a continuous state of flux and changing equilibrium. Modern international regulatory financial systems must stay in a similar state and regulate the system's equilibrium dynamically.

The traditional, functional, rigid regulatory models based on certain utopian behavioral assumptions of agent-agent relationships do not provide the necessary basis for an effective regulatory policy formulation in modern financial systems.

The challenge is to design appropriate models to regulate a complex, adaptive, diversified financial system dynamically. The uncertainties inherent in such systems are then contained, and the intended objectives in a sustainable, continuous, and resilient manner are achieved.

CAFS management approach can lead to new and emergent products, services, markets, and ventures due to resulting innovations, providing the adaptability, stability, continuity, and needed benefits of robustness, resiliency, and survivability in a dynamically changing global financial environment.

When financial systems evolve and become more complex and adaptive, the formulated regulatory policies must adapt and evolve accordingly.

Questions

Question 1. If in example 2 of this section, the probability distribution of the five-year loan rates for all banks had equal coefficient b, average μ and standard deviations of σ or:

$$P(x) = bx_0 e^{-0.5\left(\frac{x-\mu}{\sigma}\right)^2} \quad \text{and} \quad \sum_{x=0}^{x=x} bx_0 e^{-0.5\left(\frac{x-\mu}{\sigma}\right)^2} = 1$$

Find out the Average System Information Entropy or ASIP for the five-year average loan rate.

Answer 1:
We get,

- **ASIP equals to:**

$$S_1 = -\sum_{x=0}^{x=x} (bx_0 e^{-0.5\left(\frac{x-\mu}{\sigma}\right)^2}) \ln (bx_0 e^{-0.5\left(\frac{x-\mu}{\sigma}\right)^2})$$

$$S_2 = -\sum_{x=0}^{x=x} (bx_0 e^{-0.5\left(\frac{x-\mu}{\sigma}\right)^2}) \ln (bx_0 e^{-0.5\left(\frac{x-\mu}{\sigma}\right)^2})$$

. . . .

$$S_{23} = -\sum_{x=0}^{x=x} (bx_0 e^{-0.5\left(\frac{x-\mu}{\sigma}\right)^2}) \ln (bx_0 e^{-0.5\left(\frac{x-\mu}{\sigma}\right)^2})$$

$$ASIP = \frac{S_1 + S_2 + \cdots + S_{20}}{20}$$

$$ASIP = -\sum_{x=0}^{x=x} \left(bx_0 e^{-0.5\left(\frac{x-\mu}{\sigma}\right)^2} \right) \ln \left(bx_0 e^{-0.5\left(\frac{x-\mu}{\sigma}\right)^2} \right)$$

Question 2. In Question 1, as σ increases (volatility or uncertainty in the five-year average loan rate), what will happen to ASIP and ASIG.

Answer 2:

As σ increases, $e^{-0.5\left(\frac{x-\mu}{\sigma}\right)^2}$ increases which result in an increase in ASIP or the information uncertainty. It will also cause ASIG to decrease. As σ or volatility and uncertainty in the five-year average loan rate goes up, less structural certainty prevails for each bank's loan rates.

Question 3. In Question 1, as inflation or ε moves from zero to a substantial (infinity) rate, what happens to the 5YALR. Briefly analyze the result.

$$5YALR = \left(\sum_{i=1}^{i=20} x_0 b_i (1 + e^{-\varepsilon})^{-1}\right)/20$$

Answer 3:

The results show,

$$\lim_{\varepsilon \to 0} 5YALR = \lim_{\varepsilon \to 0}\left(\sum_{i=1}^{i=20} x_0 b_i (1 + e^{-\varepsilon})^{-1}\right)/20 = x_0\left(\sum_{i=1}^{i=20} b_i\right)/40$$

$$\lim_{\varepsilon \to \infty} 5YALR = \lim_{\varepsilon \to \infty}\left(\sum_{i=1}^{i=20} x_0 b_i (1 + e^{-\varepsilon})^{-1}\right)/20 = x_0\left(\sum_{i=1}^{i=20} b_i\right)/20$$

As inflation rate ε moves from zero to huge numbers, 5YALR for the 20 banks will double and grow from $x_0(\sum_{i=1}^{i=20} b_i)/40$ and maximize to $x_0(\sum_{i=1}^{i=20} b_i)/20$. The coefficient b_i for each bank represents their impact on the average rate calculations.

Question 4. Insurance companies are in the business of uncertainty reduction or risk coverage. By collecting annual premiums, pooling, and managing funds, they cover various potential future losses caused by health, life, property events for individuals and companies. Compare the general levels of management complexity of a life insurance company and a bank using the measurement parameter table (use your best judgments for relative rankings). Assume that both institutions are well managed and function in a diverse economy with 500 employees and 100,000 clients. Agents in both institutions mean personnel and clients.

Answer 4:

The results based on best judgment are,

Complex Adaptive Financial System
Bank vs. Insurance Company
"Relative Measurement Parameters"

Parameter	Code	Wt.	Bank	Insurance Company
CAS Type	CT	-	Blended (Physical, Information, Logical)	Blended (Physical, Information, Logical)
Number of Agents	NOA	1	Very High (10)	Very High (10)
Degree of Non-homogeneity of Agents	DNHA	3	High (7)	Medium (5)
Average Number of Connections per Agents	ANCA	2	Medium (5)	Medium (5)
Average Number of In Connections per Agent	ANICA	1	Medium (5)	Medium (5)
Average Number of Out Connections per Agent	ANOCA	1	Medium (5)	Medium (5)
Agent Coupling Strengths	ACS	2	High (7)	Medium (5)
Number of Hamiltonian Cycles	NHC	3	Medium (5)	Medium (5)
Average System Information Entropy	ASIP	-1	Low (3)	Medium (5)
Average System Information Energy	ASIG	2	High (7)	Medium (5)
System Dynamics	SD	2	Complex (7)	Simple (5)
Degree of Contagion Effect	DCE	3	High (7)	High (7)
Degree of Collective Intelligence	DCI	3	Medium (5)	Medium (4)
Degree of Emergence & Innovation	DEI	3	Low (3)	Low (3)
Degree of Resilience	DRE	1	Medium (5)	Medium (5)
Degree of Robustness	DRO	1	Medium (5)	Medium (5)
Weighted Average Degree of CAS	WADC		Medium (5.9)	Medium (5.1)

The WADC for the bank is 5.9 (Medium), and for the insurance company is 5.2 (Medium). As institutions, we have determined that both have medium-level management complexity, with the bank being slightly more complex. Given that the insurance company was focused on life insurance products only and the bank is more diversified in its services, its higher complexity levels make sense. If we had analyzed a cluster of banks (as we did for 20 banks in example 1 in this section) and a cluster of insurance companies (say 20 insurance companies in an economy), we would have certainly seen higher WADC for both. It will be due to the higher levels of complexity for having clusters of institutions coupled, interconnected, and interacting.

Question 5. Securities Brokerage Companies are financial intermediaries involved in brokering the trade, sale, and purchase of financial instruments, primarily for other individuals, institutions, and at times for themselves. In comparison, Investment Companies are financial intermediaries principally involved in managing individual or institutional funds for securities investments. Compare the general levels of management complexity for these two types of financial institutions using the measurement parameter table (use your best judgments for relative rankings). Assume that both institutions are well managed and function in a diverse economy with 100 employees and 10,000 clients. Agents in both institutions mean personnel and clients.

Answer 5:
The results, based on best judgment, are:

Complex Adaptive Financial System
Securities Brokerage vs. Investment Company
"Relative Measurement Parameters"

Parameter	Code	Wt.	Securities Brokerage	Investment Company
CAS Type	CT	-	Blended (Physical, Information, Logical)	Blended (Physical, Information, Logical)
Number of Agents	NOA	1	High (7)	High (7)
Degree of Non-homogeneity of Agents	DNHA	3	High (7)	Medium (5)
Average Number of Connections per Agents	ANCA	2	Medium (4)	Medium (4)
Average Number of In Connections per Agent	ANICA	1	Medium (4)	Medium (4)
Average Number of Out Connections per Agent	ANOCA	1	Medium (4)	Medium (4)
Agent Coupling Strengths	ACS	2	Medium (4)	Medium (4)
Number of Hamiltonian Cycles	NHC	3	Medium (4)	Medium (4)
Average System Information Entropy	ASIP	-1	Medium (4)	Medium (4)
Average System Information Energy	ASIG	2	Medium (5)	Medium (4)
System Dynamics	SD	2	Complex (7)	Simple (6)
Degree of Contagion Effect	DCE	3	High (7)	Medium (5)
Degree of Collective Intelligence	DCI	3	Medium (5)	Medium (5)
Degree of Emergence & Innovation	DEI	3	Medium (5)	Medium (5)

Degree of Resilience	DRE	1	Medium (5)	Medium (5)
Degree of Robustness	DRO	1	High (7)	High (7)
Weighted Average Degree of CAS	WADC		Medium (5.4)	Medium (4.9)

The relative WADC for the securities brokerage company is 5.4 (Medium), and for the investment company is 4.9 (Medium). As institutions, they both have medium-level management complexity, with the securities brokerage company being slightly more complex. If we had analyzed a cluster of securities brokerage companies or their industry versus the investment company industry, we would see higher WADC for both due to the higher levels of complexity for having a group of interconnected and interacting institutions in an industry.

Question 6. An investment company is a financial intermediary involved in managing individual or institutional funds for investments into securities. They buy and sell securities while creating a securities portfolio to maximize return and reduce risk with various objectives such as constant income or growth in value and even combining the two. Let us assume that we have an investment company with a diversified portfolio of 60 securities with minimum associated risks and uncertainties. The monthly returns for individual securities in this portfolio have various correlations between securities[52] (various couplings). The portfolio is being managed daily for the generation of the maximum total monthly return. Assuming that each security in the portfolio is an agent with nonlinear behavior and interactions with other securities, determine the investment portfolio's network classification, attributes, and interaction coupling strength matrix (remember that 60 stocks have return correlations).

---

[52] There is a difference between Correlation and Causation. When we have an observed relationship or pattern between two variables, agents or items, we are observing Correlation. Correlation simply shows a relationship determined through various statistical observations and can be positive, negative, linear or nonlinear. When we have one variable, agent or item causing another variable, agent or item we observe Causation. Causation and Correlation can exist at the same time, but Correlation between two variables, agents or items does not mean they cause one another. Correlation does not mean Causation, since Correlation could have been caused by a third variable, agent or item. Causation will result in Correlational observations if we have enough number of statistically significant observations. If we have very few observations for a cause and effect phenomena, Causation is there but no Correlation can be detected due to the lack of enough data.

Assume that the total portfolio weighted average monthly net return is x. This is the monthly net portfolio return after consolidating the 60 security returns considering their corresponding weights in the portfolio, correlational effects, and trading and management fees. It has been determined that x through time follows a pattern based on the average monthly inflation rate ε and average monthly 1year short term treasury rates δ as follows:

$$x(t) = \frac{\varepsilon + \delta}{(1 + e^{-t})}$$

Answer 6:
We have,

- **System Graph:** Not shown. Imagine a cluster of 60 agents all connected
- **CAS Type:** Information
- **System Attributes:**

✓	NOA	60 securities or agents
✓	DNHA	Non-homogeneous
✓	ANCA	3600 (60 x 60) possible links with an avg. of 60 per security. These links determine the return correlational effects of each security on others. We are assuming that the monthly return of each security can affect its own returns.
✓	ANICA	3600 links with an avg. 60 per security
✓	ANOCA	3600 links with an avg. 60 per security
✓	ACS	Defined as $c_{ij}(t)$ not shown
✓	NHC	Many
✓	ASIP, ASIG	Due to the return correlational impact of each security on others and fluctuations due to market information on a daily basis, the average system information entropy or ASIP will be high (due to fluctuations in returns) and the average system information energy or ASIG will be low.
✓	SD	Very complex. This is a complex adaptive financial system with 60 monthly security returns all correlated with various feedback and feedforward influences. Securities (agents) trigger parameters (monthly returns) represent a vector with 60 elements of $T_1(t)$ to $T_{60}(t)$ and a 60 x 60 interaction coupling

matrix that results in the total portfolio monthly return x as follows:

$$T(t) = \begin{pmatrix} T_1(t) \\ \vdots \\ T_{60}(t) \end{pmatrix}$$

$$C(t) = \begin{pmatrix} c_{11}(t) & \cdots & c_{1\,60}(t) \\ \vdots & \ddots & \vdots \\ c_{60\,1}(t) & \cdots & c_{60\,60}(t) \end{pmatrix}$$

Due to the daily trading of the 60 securities in the portfolio, the monthly return of each security and the interaction coupling matrix can mutually and dynamically affect one another and change through time. Again we will observe a co-interacting, co-evolving adaptive behavior between securities and their respective interactions. Mathematically this can be expressed in terms of two sets of 60 coupled nonlinear partial differential equations. The general mathematical form for this system of equations can be expressed as:

$$\frac{dT_i(t)}{dt} = R\left(c_{ij}(t), T_j(t)\right) \quad for\ i,j\ from\ 1\ to\ 60$$

$$\frac{dc_{ij}(t)}{dt} = W\left(c_{ij}(t), T_j(t)\right) \quad for\ i,j\ from\ 1\ to\ 60$$

Functions R and W are different for each security.

- **Input/Source:** ε (Avg., monthly inflation), δ (Avg. monthly 1 year short term treasury rate), and time t.

- **Output/Sink Result:** $x(t) = \frac{\varepsilon + \delta}{(1 + e^{-t})}$

- **Interaction Coupling Strength Matrix:**

$$C(t) = \begin{pmatrix} c_{11}(t) & \cdots & c_{1\,60}(t) \\ \vdots & \ddots & \vdots \\ c_{60\,1}(t) & \cdots & c_{60\,60}(t) \end{pmatrix}$$

This matrix changes dynamically through time, and the change can mathematically be represented as:

$$\frac{dC(t)}{dt} = \begin{pmatrix} \frac{dc_{11}}{dt} & \cdots & \frac{dc_{1\,60}}{dt} \\ \vdots & \ddots & \vdots \\ \frac{dc_{60\,1}}{dt} & \cdots & \frac{dc_{60\,60}}{dt} \end{pmatrix}$$

Question 7. Determine in the previous question, as time t grows from zero, what happens to x given the following observed relationship. Briefly analyze the result.

$$x(t) = \frac{\varepsilon + \delta}{(1 + e^{-t})}$$

Answer 7:

The results show,

$$\lim_{t \to 0} x = \lim_{t \to 0} \left(\frac{\varepsilon + \delta}{(1 + e^{-t})} \right) = \frac{\varepsilon + \delta}{2}$$

$$\lim_{t \to \infty} x = \lim_{t \to \infty} \left(\frac{\varepsilon + \delta}{(1 + e^{-t})} \right) = (\varepsilon + \delta)$$

As time t moves from zero to huge numbers, x or average total monthly portfolio return stays in a corridor and will double from $\frac{\varepsilon+\delta}{2}$ to a maximum of $(\varepsilon + \delta)$. Interestingly enough, the maximum monthly return of the portfolio, in the long run, will be equal to the average monthly inflation rate ε plus the average monthly 1year short-term treasury rate δ. It is better to keep this investment for the long term.

Question 8. If in the previous question, the probability distribution of the monthly return of the portfolio was equal to $P(x)$ as follows:

$$P(x) = \lambda e^{-0.5\left(\frac{x-\mu}{\sigma}\right)^2} \quad \text{and} \quad \sum_{x=0}^{x=x} \lambda e^{-0.5\left(\frac{x-\mu}{\sigma}\right)^2} = 1$$

With constant λ and average of μ and standard deviation of σ for the portfolio, find the Average System Information Entropy or ASIP for the return.

Answer 8:
We get,

- **ASIP equals to;**

$$ASIP = -\sum_{x=0}^{x=x}(\lambda e^{-0.5\left(\frac{x-\mu}{\sigma}\right)^2})\ln \lambda e^{-0.5\left(\frac{x-\mu}{\sigma}\right)^2}$$

Question 9. If, in the previous question, the standard deviation of σ doubles, what happens to the probability distribution of the monthly return of the portfolio.

Answer 9:
We have,

$$P_\sigma(x) = \lambda e^{-0.5\left(\frac{x-\mu}{\sigma}\right)^2}$$

Moreover, if σ becomes 2σ, we get:

$$P_{2\sigma}(x) = \lambda e^{-0.5\left(\frac{x-\mu}{2\sigma}\right)^2} = \lambda e^{(-0.5\left(\frac{x-\mu}{\sigma}\right)^2)/4}$$

$P_{2\sigma}(x)$ is larger than $P_\sigma(x)$

It means that the distribution will become wider, and the uncertainty levels (risks) will increase as σ is doubled.

Question 10. The monthly return on two open-end funds (mutual funds) are the results of a complex adaptive dynamic and resulted in observations of R_i and σ_i in various samples i in the market. The following table shows all the observations made. Calculate the Hurst exponents for each fund performance? Which output could potentially behave more chaotically?

Calculation of the Hurst Exponent for two open-end funds									
R_i^1	σ_i^1	$\ln(\frac{R_i^1}{\sigma_i^1})$	R_i^2	σ_i^2	$\ln(\frac{R_i^2}{\sigma_i^2})$	i	$\ln i$	H_i^1	H_i^2
20	5.50		15	3.50		10			
23	5.00		17	2.50		20			
27	5.00		22	2.20		30			
32	5.00		25	2.00		40			
36	5.00		25	1.60		50			
42	5.50		27	1.40		60			
50	6.00		30	1.30		70			
55	6.00		32	1.10		80			
62	6.50		34	1.00		90			
65	6.50		38	1.00		100			

Answer 10:
The following table shows the results.

Calculation of the Hurst Exponent for two open-end funds									
R_i^1	σ_i^1	$\ln(\frac{R_i^1}{\sigma_i^1})$	R_i^2	σ_i^2	$\ln(\frac{R_i^2}{\sigma_i^2})$	i	$\ln i$	H_i^1	H_i^2
20	5.50	1.29	15	3.50	1.46	10	2.30	0.56	0.63
23	5.00	1.53	17	2.50	1.92	20	3.00	0.51	0.64
27	5.00	1.69	22	2.20	2.30	30	3.40	0.50	0.68
32	5.00	1.86	25	2.00	2.53	40	3.69	0.50	0.68
36	5.00	1.97	25	1.60	2.75	50	3.91	0.50	0.70
42	5.50	2.03	27	1.40	2.96	60	4.09	0.50	0.72
50	6.00	2.12	30	1.30	3.14	70	4.25	0.50	0.74
55	6.00	2.22	32	1.10	3.37	80	4.38	0.51	0.77
62	6.50	2.26	34	1.00	3.53	90	4.50	0.50	0.78
65	6.50	2.30	38	1.00	3.64	100	4.61	0.50	0.79

As the sample sizes grow from 10 to 100, the first mutual fund's monthly return Hurst exponent stays relatively constant in the range of 0.56 to 0.50 (a random walk behavior or pure noise). On the contrary, the second mutual fund's monthly return Hurst exponent moves from 0.63 (smaller persistence and black noise), showing a relative sign of chaotic tendencies towards 0.79 (a larger persistence and higher black noise) with near chaotic dynamics. The second output could potentially reach chaotic behavior faster and is considered a riskier investment.

Question 11. Through time-series studies, it has been determined that the monthly return of a Unit Investment Trusts or UIT (a form of investment company) through time t is the results of very complex adaptive financial dynamics and follows a pattern based on the average monthly inflation rate ε and the average monthly return on the Dow Jones Industrial Average ψ, as follows:

$$x(t) = \varepsilon + \frac{1.5\psi}{(1 + e^{-t})}$$

As time t grows from zero, what happens to x. Analyze the result.

Answer 11:
The results show,

$$\lim_{t \to 0} x = \lim_{t \to 0} \left(\varepsilon + \frac{1.5\psi}{(1 + e^{-t})} \right) = \varepsilon + \frac{1.5\psi}{2}$$

$$\lim_{t \to \infty} x = \lim_{t \to \infty} \left(\varepsilon + \frac{1.5\psi}{(1 + e^{-t})} \right) = \varepsilon + 1.5\psi$$

As time t grows, x or average total monthly return of UIT stays in a corridor of minimum equal to $(\varepsilon + \frac{1.5\psi}{2})$ to a maximum of $(\varepsilon + 1.5\psi)$. The maximum monthly return of the UIT, in the long run, will be equal to the average monthly inflation rate ε plus 1.5 times the average monthly return on the Dow Jones Industrial Average ψ. It is a long-term investment because, at more prolonged periods, it covers average inflation rates and provides an additional return over and above the returns of the Dow Jones Industrial Average.

Question 12. If in the previous question, the probability distribution of the monthly return of the UIT was equal to $P(x)$ as follows:

$$P(x) = e^{-\left(\frac{x}{\sigma}\right)^2} \text{ and } \sum_{x=0}^{x=x} e^{-\left(\frac{x}{\sigma}\right)^2} = 1$$

With a constant standard deviation of σ for the UIT return, find out the Average System Information Entropy or ASIP of the UIT return.

Answer 12:
We get,

- ASIP equals to:

$$ASIP = -\sum_{x=0}^{x=x} e^{-\left(\frac{x}{\sigma}\right)^2} \ln e^{-\left(\frac{x}{\sigma}\right)^2}$$

$$ASIP = \sum_{x=0}^{x=x} \left(\frac{x}{\sigma}\right)^2 e^{-\left(\frac{x}{\sigma}\right)^2}$$

Question 13. In the previous question, what happens to the Average System Information Entropy or ASIP of the UIT return if the standard deviation σ doubles.

Answer 13:
We have,

$$ASIP_\sigma(x) = \left(\sum_{x=0}^{x=x} \left(\frac{x}{\sigma}\right)^2 e^{-\left(\frac{x}{\sigma}\right)^2}\right)$$

$$ASIP_{2\sigma}(x) = \left(\sum_{x=0}^{x=x} \left(\frac{x}{2\sigma}\right)^2 e^{-\left(\frac{x}{2\sigma}\right)^2}\right) = \left(\frac{1}{4}\right)\left(\sum_{x=0}^{x=x} \left(\frac{x}{\sigma}\right)^2 e^{-\left(\frac{1}{4}\right)\left(\frac{x}{\sigma}\right)^2}\right)$$

$$ASIP_{2\sigma}(x) \text{ is larger than } ASIP_\sigma(x)$$

If the standard deviation σ doubles, ASIP will become larger. Intuitively this makes sense. As the standard deviation or the volatility of the UIT's monthly returns doubles, the Average System Entropy or uncertainty of returns increases.

Question 14. In the previous question, what happens to the Average System Information Entropy or ASIP of the UIT return if the standard deviation σ moves from zero to a huge number. Analyze the results briefly.

Answer 14:

We have,

$$ASIP = \left(\sum_{x=0}^{x=x} \left(\frac{x}{\sigma}\right)^2 e^{-\left(\frac{x}{\sigma}\right)^2}\right)$$

$$\lim_{\sigma \to 0} ASIP = \lim_{\sigma \to 0} \left(\sum_{x=0}^{x=x} \left(\frac{x}{\sigma}\right)^2 e^{-\left(\frac{x}{\sigma}\right)^2}\right) = 0$$

$$\lim_{\sigma \to \infty} ASIP = \lim_{\sigma \to \infty} \left(\sum_{x=0}^{x=x} \left(\frac{x}{\sigma}\right)^2 e^{-\left(\frac{x}{\sigma}\right)^2}\right) = 0$$

It means that if the standard deviation of monthly return σ starts from zero and gets very large, ASIP will become zero at both extremes. There will be a value of the standard deviation of return σ at which ASIP will be maximized. That value is reached at $x = \sigma$ at which ASIP is maximized and equals to:

$$ASIP_{Max} = \left(\sum_{x=0}^{x=x} (1)^2 e^{-(1)^2}\right) = \frac{x+1}{e}$$

Question 15. If the monthly returns for two Unit Investment Trusts (UITs) show nonlinear CAS behavior and for the $\delta(t_i)$ and $\delta(t_{i-1})$ observations, we get the following table. Calculate the two total Lyapunov exponents? How do we interpret the behavior of the returns?

\multicolumn{9}{c}{Calculation of the Lyapunov Exponent for monthly returns of two UITs}
$\delta(t_i)$ 1	$\delta(t_{i-1})$ 1	$\ln(\frac{\delta(t_i)\ 1}{\delta(t_{i-1})\ 1})$	$\delta(t_i)$ 2	$\delta(t_{i-1})$ 2	$\ln(\frac{\delta(t_i)\ 2}{\delta(t_{i-1})\ 2})$	t	$\frac{1}{t}$	λ_t^1	λ_t^2
0.010	0.011		0.050	0.040		10			
0.012	0.014		0.044	0.036		10			
0.014	0.016		0.040	0.032		10			
0.015	0.017		0.037	0.030		10			
0.016	0.018		0.032	0.026		10			
0.018	0.020		0.025	0.020		10			
0.020	0.022		0.022	0.018		10			
0.023	0.025		0.020	0.016		10			
0.025	0.028		0.027	0.022		10			
0.028	0.031		0.018	0.014		10			
Sum λ^1			Sum λ^2						

Answer 15:

The following table shows the results.

Calculation of the Lyapunov Exponent for monthly returns of two UITs									
$\delta(t_i)^1$	$\delta(t_{i-1})^1$	$\ln(\frac{\delta(t_i)^1}{\delta(t_{i-1})^1})$	$\delta(t_i)^2$	$\delta(t_{i-1})^2$	$\ln(\frac{\delta(t_i)^2}{\delta(t_{i-1})^2})$	t	$\frac{1}{t}$	λ_t^1	λ_t^2
0.010	0.011	-0.095	0.050	0.040	0.223	10	0.1	-0.010	0.022
0.012	0.014	-0.154	0.044	0.036	0.201	10	0.1	-0.015	0.020
0.014	0.016	-0.134	0.040	0.032	0.223	10	0.1	-0.013	0.022
0.015	0.017	-0.125	0.037	0.030	0.210	10	0.1	-0.013	0.021
0.016	0.018	-0.118	0.032	0.026	0.208	10	0.1	-0.012	0.021
0.018	0.020	-0.105	0.025	0.020	0.223	10	0.1	-0.011	0.022
0.020	0.022	-0.095	0.022	0.018	0.201	10	0.1	-0.010	0.020
0.023	0.025	-0.083	0.020	0.016	0.223	10	0.1	-0.008	0.022
0.025	0.028	-0.113	0.027	0.022	0.205	10	0.1	-0.011	0.020
0.028	0.031	-0.102	0.018	0.014	0.251	10	0.1	-0.010	0.025
Sum		-1.125	Sum		2.167				
λ^1		-0.113	λ^2		0.217				

The total Lyapunov exponent for the first UIT's monthly return is negative and equal to -0.113, and for the second one, positive and equal to 0.217. Both Lyapunov exponents stay relatively constant. The first one stays in the range of -0.008 to -0.015, and the second one in the range of 0.020 to 0.025. The first UIT Lyapunov exponent is negative and shows a tendency towards a stable monthly return over time. The second UIT Lyapunov exponent is positive and shows a potential tendency towards instability and monthly returns' chaotic behavior. The first UIT has lower risks due to a movement towards more stability of return over time.

Final Word

All CAS structures include network-based nonlinear interactions of many cognitive agents. In this book, we first explained what a network is and analyzed it. To do this analysis properly, we had to go over the concept of graphs which was crucial in understanding the interconnection and interactions among elements/agents in a network. We next reviewed the basic principles of systems and looked at several types of system dynamic equations and how agent-agent interactions can behave and result in specific outputs that move towards chaotic dynamics. We next applied the mathematical techniques from systems and networks to quantify the behavior of Complex Adaptive Systems. We finally applied all the reviewed concepts to the financial systems by defining these systems as Complex Adaptive Financial Systems or CAFS. Through examples, we saw how one could benefit from a CAS viewpoint in such complex financial systems. Analyzing and comparing current socio-political and economic issues such as Globalization, Nationalization, Privatization, and Patriotism, Cultism can also be looked at with a CAS prism creating interesting results and recommendations.

Throughout the book, we tried to exclude theoretical and mathematical concepts and emphasized more applied and conceptual issues related to CAS. Some quantitative analysis was required, and we tried to eliminate detailed and complex mathematical derivations and calculations.

This book is an introductory view on how to quantify specific properties of Complex Adaptive Systems. It should be helpful for practitioners that have been exposed to qualitative and quantitative modeling concepts with a basic knowledge of statistics, calculus, differential equations, network, and system theory.

In CAS, we have independent, changing components with non-linear, dynamic, and adaptive interactions, making the components and interactions highly unpredictable. The trajectory and dynamics of the system are complicated.

Simply put, nonlinear interactions between agents cause complexity, and learning plus cognition causes adaptive behavior.

Definitions[53]

Adaptation

In a complex system, adaptation, also known as homeostasis, is a process by which, through experience and memory, a learning mechanism guides change in the system's structure. As time passes, the system makes better use of its environment to reach robustness, resiliency, and continuous stability. It usually happens through relationships between agents through contagion, collaborative, coordinated, cooperative, coupling, synchronized, self-control, and regulating algorithms and processes.

In a biological context, adaptation is a trait that increases an individual's or group's fitness in a particular environment. The process of adaptation occurs via modifications of the behavior of an individual or group. Homeostasis is defined as an organism's ability or dynamic system to maintain a stable internal environment, even as its external environment changes. The Greek word "homeo" means similar, and "stasis" means stable, and together they mean continuous stability. This is often important for regulating the system's continued function and usually employs balancing feedback mechanisms. The system's rigidity to change can also be defined as the degree of Homeostasis.

Collective or Swarm Intelligence

When specific cognitive agents or individuals with no central controls group together while following simple rules and algorithms, they gain a collective behavior, better known as "Swarm Intelligence." ***It is a holistic, synergistic, distributive, and coordinated intelligence capable***

---

[53] With some changes, similar to the Appendix of the book; **Shayan, S. A. (2019)**,"Understanding Complex Adaptive Systems," Independent Publisher, Amazon.

of solving complex problems, which would be difficult for an isolated agent or individual to solve. Research has shown that simulated Swarm Intelligence-based decision-making processes can improve accuracy up to forty percent while generating more innovative, emergent behavior and solutions. It is preferable to use a collection of cognitive individuals in committee or council-based setups, with clarified rules and mandates (or algorithms), to make complex problems to reduce the possible errors and risks of needed decisions. *More Swarm Intelligence means more memory, order, emergent behavior, and complexity in the system*.

Contagion effect

The contagion effect happens when coupled, synchronized, coordinated, modulated, collaborative, cooperative, and synergistic local interactions among elements, parts, pieces, or agents exist. It can give rise to the formation of order and disorder at the system's macrostructures. It is a term primarily used in disease prevention management and virus growth and mutation. It helps understand how agent-agent interactions can be coupled, synchronized, and spread an external shock or perturbation throughout the system.

Complexity

Complexity means different things in different disciplines. When a system is complex, it is difficult to understand and analyze its dynamic behavior. In general, the complexity of a system emerges from the interactions of its interrelated elements, components, parts, pieces that we call agents, as opposed to the characteristics of those agents in and of themselves. Agents can have reactive or proactive (be cognitive and think) behaviors. Complexity science is studying nonlinear and emergent system behavior and seeks to understand how its complex behavior arises from its interacting parts or agents. Generally speaking, due to the nonlinearity of agent interactions, complex behavior cannot be reduced to or derived from the sum of the behavior of the system's agents. Complexity is a new approach to study the relationship between agents of a system and how they can lead to holistic, emergent, synergistic, robust, and resilient collective behavior.

Emergence

Emergence or Saltation is when a structured aggregate behavior in a CAS arises from localized agent behavior. It is a holistic process by which a CAS of interacting subunits or agents suddenly acquires new properties that cannot be understood through the simple addition of the agent contributions. *The concept of emergence is the formation of order through synchronized, collective, and modulated local interactions of parts, pieces, or agents in a bottom-up and top-down manner that gives rise to a holistic sudden macro organizational structure.* Emergence is generally considered a process that leads to the appearance of structure not directly described by the defining constraints and forces that control the CAS. In emergence, the localized behavior of agents aggregates into global behavior that is, in some sense, disconnected from its origin. There can be multiple layers of emergence. One can observe emergence at lower or agent-agent levels or higher agent cluster levels. More heterogeneity or diversity of agents in CAS leads to more flexibility and the possibility of observed emergent and new agent-agent interactions.

Emergence or Saltation is not imposed by a central force or management process but results from an interactive rule-based structure between agents, affecting the system at various levels. Over time, emergent behavior is directly related to innovation, leading to evolution, new structures, resiliency, robustness, and adaptable systems.

There is a difference between learning and innovation or emergence. In learning, we adapt to recognize an existing set of patterns, usually spelled out by the environment. With innovation, very similar to emergence and saltation, we may suddenly view things differently. Saltation or the theory of volatile evolution is a viable theory of evolution that can describe Complex Adaptive Systems' innovation phenomena. *The more complex a system becomes, the more emergent behavior it will have.*

Entropy

Entropy can be considered as the natural tendency of a closed system (a system with no external interactions) to move from a more ordered (or with less structural flexibility, less degrees of freedom, and less symmetry) to a less ordered (or with more structural flexibility, more degrees of freedom and more symmetry) state. *According to the second law of thermodynamics, closed systems, including the universe, tend to degrade or lose their structures in the absence of energy inputs outside of the system, leading to an increase of a property called entropy or structural randomness.* Therefore the macro entropy of the universe is continually increasing. In the language of statistical mechanics, the definition of Entropy refers to the number of possible microstates corresponding to a given macro-state. ***In a closed system, as Entropy increases, structural flexibility, degrees of freedom, and symmetry would increase, but complexity decreases.***

In information theory, entropy measures the distribution of discrete states in a system. From an information theory perspective, a uniform and homogenous distribution would have maximum entropy. In modern interpretations, Entropy is the amount of additional information needed to specify the exact physical state of a system. Understanding entropy's role in various processes requires understanding how and why the information changes as the system evolve from its initial to its final condition. Less Information, more structural flexibility, or degrees of freedom and symmetry are directly related to higher Entropy measures. ***As Entropy is reduced, more complexity is gained.***

Evolution

It is a process observed in all dynamics adaptive systems such as biological, sociological, cultural, political, and economic systems, to name a few. It is when new elements, agents, parts, internal interactions, and structures are created from within the system and become a part of the system. Evolution can be destructive or creative and cause an increase or decrease in the diversity of the system. Evolutionary dynamics can be statistical or non-statistical and seems to be universal characteristics of all complex systems. There are definite relationships between diversity, robustness, resiliency, adaptability, emergence, and evolution of a system. In evolutionary Complex Adaptive Systems, the boundaries change and evolve through time.

Heterogeneity

Heterogeneity or diversity of agents is an important feature of complex adaptive systems. We still do not understand how heterogeneity influences the holistic behavior of complex systems or networks. We know that it allows systems to have different types of agents with distinct behavior, which leads to diversity. If we have homogeneous agent structures, system behavior can be analyzed using average laws and behaviors with potential synchronized and contagion effects leading to wild fluctuations and possible system collapse. Proper heterogeneity or diversity of agents can lead to continuous agent-agent canceling effects, adjustments, self-correcting, and stabilizing mechanisms in the system. It can also prevent the system from having wild fluctuations and volatilities, leading to more system robustness and continuous dynamic stability.

Mutuality or Interconnections

Mutuality or the degree of interconnection and connectivity defines the number of interactions or connections each interrelated element, component, part, piece, or agent has with the other agents. The smaller the average degree of connectivity or mutuality between agents, the less complex behavior we will observe in the system. The higher the number of interactions, the more complex dynamic behavior prevails. A connectivity matrix is usually used to show the degree of mutuality, interconnection, coupling, or topology of the system (using graph theory concepts). The higher degree of mutuality can lead to more agent-agent interconnections, leading to positive or negative correlations, couplings, cooperation, collaboration, modulation, synchronization, synergies, and emergent behavior.

Nonlinear Behavior

When the system output or results cannot be determined or analyzed through a simple linear combination (adding or subtracting) of the inputs, we have a nonlinear dynamic behavior in our system. This means that the superposition principle does not work, or the whole is not the sum of its parts. All complex systems have nonlinear behavior, and their resulting outputs cannot be determined from a linear combination of their inputs. Nonlinear behaviors under certain conditions can become unpredictable, chaotic, and unstable.

Resilience

Resilience is defined as the ability of a system to persist and maintain its core functions and purpose in the presence of noise, disturbance, perturbation, stress, or other changes in its environment. A complex system's resiliency is directly related to the degree of adaptability or homeostasis of that system. The degree of resilience is enhanced when we have locally distributed interacting and dynamic controls. Therefore more central controls negatively impact the system's resilience and adaptability.

Robustness

Like resilience and adaptation, robustness is the capability of a system to go back to its equilibrium state by correcting shock, error, noise, disturbance, or perturbation created in its inputs and structures by the environment. *While being a short-term phenomenon, it means that resilience and adaptation are the long-term effects of robustness.* The more robust a system is, the quicker the correction is made and resilient, adaptive, and equilibrium state is reached. The system's robustness is a function of the time reached to equilibrium state under shocks. Another way to describe robustness is to maintain a specific behavior, trait, or characteristic regardless of changing environmental conditions. The robustness in a complex system is enhanced when we have locally distributed interacting and dynamic controls. More central control negatively impacts the system's robustness. More existing distributed controls between agents, such as negative feedback structures, lead to more system robustness. This results in a quicker distribution of shock or noise effects throughout the system and a speedier movement towards a state of equilibrium. Robustness is a valuable feature that helps a system maintain its stability and equilibrium.

Self-Organization

Self-organization or pattern formation is a process in which patterns at the global levels of a system emerge through time, solely from numerous interactions among the lower-level components of that system. Moreover, the rules specifying interactions among the system's components or agents are executed using local information without reference to the global pattern. The self-organizing behavior of complex adaptive systems happens through time, and it is related to the microscopic dynamic behavior of the system, leading to macroscopic structures. Emergence is about the scale and is the result of microscopic dynamics and self-organization. *Therefore from a holistic structural viewpoint, self-organization has an element of time and scale, while emergence has an element of space attached. Self-organization is a dynamic process that helps the understanding of the Emergent phenomena in complex adaptive systems.*

Self-Organized Criticality

Self-organized criticality refers to the notion that there can be situations where a CAS can have critical points of triggering and reshaping that will be reached spontaneously throughout its dynamics. This is where the CAS equilibrium can become unstable, highly sensitive, and responsive to shock, error, or change resulting in reduced resiliency levels and robustness. It means that the system has sensitive dependence on its initial conditions. Cases such as financial bubbles and meltdowns, social and political revolutions, avalanches, and ferromagnetic phase changes fall under such behavior.

Self-Similarity

Self-similarity is a phenomenon that occurs when the structure of a sub-system resembles the structure of the system as a whole. Also, the structure of that sub-system within the original sub-system resembles the structure of the larger sub-system, and so on. Self-similarity is the defining property of structures. Self-similarity happens when a system has a repeating structure, starting from micro to macro levels, like Fractal structures.

Sensitive dependence on initial conditions

A system's sensitivity to initial conditions refers to the role of the system's starting configuration in determining the subsequent states. When this sensitivity is significant, slight changes in starting conditions or positions will significantly affect different equilibrium positions and conditions in the future. Sensitive dependence on initial conditions leads to instability, unpredictability, and chaos in nonlinear dynamical systems theory similar to positive values of λ (Lyapunov exponent) or a value close to one for H (Hurst exponent). When an error in inputs leads to considerable error in a system's long-term behavior, we have a high sensitivity to initial conditions. *It is similar to having a system with low resiliency, adaptation, and robustness. This condition happens mainly in systems with nonlinear and complex interactions between elements, components, parts, pieces, or agents enhanced through feedback and feedforward effects that exponentially reinforce the error effect in the long-term outputs. Low sensitivity to initial conditions in a complex system is considered a positive feature since it leads to a more long-term robust and stable behavior*. The more positive feedback and nonlinear interactions between the system's elements, components, parts, pieces, or agents, the more sensitivity to initial conditions we will observe. Positive feedbacks between agents or components will create amplifying effects, and negative feedbacks will create dampening effects.

More negative feedbacks among agents are preferable because it can lead to lower sensitivities to initial conditions.

Statistical Mechanics

The field of statistical mechanics seeks to explain how macroscopic behaviors emerge from the statistical properties of large numbers of "microscopic" components making up the system. There are usually many microstates that can lead to one and defined macro-state. Statistical mechanics measures the probabilities for various possible microstates that can lead to the defined macro-state. Ludwig Boltzmann originally developed statistical mechanics as a foundation for thermodynamics. However, statistical mechanics has since been applied in many fields, ranging from physics to social sciences. Suppose the microstates are quantum mechanical, such as a large number of interacting atoms or molecules. In that case, we apply the statistical mechanics' techniques to quantum mechanical bodies, or we use statistical quantum mechanics.

Stochastic process

A stochastic process whose behavior (i.e., transition from one state to a successor state) has random or probabilistic components. A classic example is a random walk process. A random walk is a simple stochastic or probabilistic process. Black-Scholes option pricing or financial diffusion equations are examples of complex stochastic partial differential equations describing the interconnected system of probabilistic financial variables.

Structural Degree of Freedom

In a CAS, the structural degree of freedom will be the number of ways in which the interacting elements, components, parts, pieces, or agents can be replaced without changing the properties or dynamic behavior of the system. It is very much related to the definition of symmetry. *Therefore the degree of freedom of a CAS is directly related to its degree of symmetry. The more rigid a CAS structure is concerning agent replacements (meaning that if you change an agent, the structure and system behavior is not sensitive and does not change dramatically), the less degree of freedom or symmetry it has. More structural flexibility and agent homogeneity in any CAS lead to a higher degree of freedom and symmetry. More system structure can lead to more system memory, causing less symmetry, less entropy, and more complexity.*

Structures & Symmetry

In any system, existing controls, structures, symmetries, and organizational patterns are always caused by local interaction of interrelated elements, components, parts, pieces, or agents and not through a centrally driven forced structure. Lack of central controls and having locally distributed interacting and dynamic controls is always an essential but insufficient feature for having a CAS behavior. As in large networks, we can observe communities within extensive agent-agent interactions. By community, we mean a set of agents at which each agent's connections in the set mostly lead to other agents in that set. The more agent-agent connections exist inside each set, the more specialized or localized this community or cluster of agents will be. Like large networks, the agent-agent connections can have positive or negative feedback and a feedforward loop structure, leading to a more complex hierarchical CAS structure with more possibilities of CAS emergent behavior.

There are also various kinds of networks. The more order and structure a network has, the less degree of freedom and symmetry will prevail. In this context, symmetry is defined as the structural indifference of a network when exposed to some form of transformation or shock. In simple terms, if a network's performance is invariant under a specific transformation, then its behavior is symmetric under that transformation. *More specialized or localized node-node communities and hierarchical structure in a network mean lower degrees of freedom, leading to less structural flexibility and, therefore, lower symmetry degrees.*

Symmetry

Symmetry can be defined in many ways. It could refer to space, time, information, or geometrical shapes. It is related to structural indifference when exposed to some form of transformation. If the system's physical structure and mathematical or informational outputs stay the same after a particular transformation, it is symmetric. *In simple terms, if a system's behavior is invariant under a particular transformation, then its behavior is symmetric under that transformation.*

For example, if we have a geometrical form such as a square and rotate it 90 degrees, we will have the same square as before in terms of shape and geometrical location. If we rotate it 120 degrees, we will have the same square but not in the exact two-dimensional location as before. A square is symmetric at multiples of 90 degrees and not symmetric at multiples of 120 degrees. We have 360 degrees for a total rotation, and therefore a square will have 4 degrees or orders of symmetries at 90 degrees. There is an infinite number of rotational symmetry for a circle in two dimensions. If we have a highly jagged physical structure, we can assume that it will have a low degree of symmetry. A highly uniform and homogeneous liquid or gas is highly symmetric by exchanging matter with a similar matter. It means if we replace parts of its matter with similar material, the properties or behavior of the liquid or gas will not change. If such a system's outputs or complexity do not change when gone through some form of transformation, then the systems are symmetric through those transformations. The fundamental laws in physics are symmetric through space and time transformations, or they are time and space invariant. It means that if we change the coordinate systems or go backward or forward through time, these laws will not change in terms of their outputs or the dynamics that they describe. Therefore they are symmetric through time and space. The concept of symmetry becomes very useful when analyzing the holistic degree of flexibility, stability, robustness, resilience, and tolerance of the structure a Complex Adaptive System has when faced with external shocks and perturbations. A one-time shock, error, or perturbation to a homogeneous structure can break or reduce the structural symmetry while the system returns to its original equilibrium. *Higher symmetry leads to higher structural flexibility or degrees of freedom, lower system memory, higher entropy, and lower complexity.*

Symmetry in CAS

Through a CAS lens, symmetry is related to the structural indifference of the system when exposed to some form of transformation. If the CAS's physical structure and mathematical or informational outputs stay the same after a particular transformation, it is symmetric.

In this context, symmetry is defined as the structural indifference of a CAS when exposed to some form of transformation. In simple terms, if a CAS's behavior is invariant under a particular transformation, then its behavior is symmetric under that transformation.

The concept of symmetry becomes very useful when analyzing the holistic degree of structural flexibility, stability, robustness, resilience, entropy, and tolerance of the structure a CAS is exhibiting when faced with external shocks or perturbations.

Synergy

Synergy in Greek means working together. Synergy is a phenomenon or process where the elements, components, parts, pieces, or cognitive or non-cognitive agents in a system will interact and cooperate advantageously for an outcome or cause with a combined effect greater than the sum of the separate effects. When extra value or result is created through the dynamic interaction of the components, we observe synergies. It happens when the whole is greater than the sum of its parts. It is a property of certain nonlinear systems with unique synergy-creating features. Synergy is positive (Positive Synergy) when the combined effect is greater than the sum of the separate effects and negative (Negative Synergy) when the combined effect is less than the linear sum of the separate effects. *Positive Synergy is desirable, and Negative Synergy is not.*

Systems

A system is a collection of interacting elements, components, parts, pieces, or, generally speaking, agents. System theory is defined as the mother of complexity theory and can be very powerful and formally abstract. Systems are usually characterized by the manner they interact with their environments. Open systems have complete interactions with their environments and are always influenced by them. They are more dynamic and complicated to understand and manage. Semi-Open systems have limited interactions with their environments and are partially influenced by the outside. They are better understood and managed. Closed systems have no interactions with their environments and have no influences from the outside. They are easier to understand and manage. All types of systems have inputs, processes, and outputs. Every system has a purpose that is revealed through its outputs. Systems can have feedback and feedforward loop structures within them, adding to their more structural complexity. Some systems are simple, have complex structures and dynamics that change their inputs, processes, outputs, and even boundaries.

System Memory

As a system or network of random, symmetric, non-interactive elements, nodes, or agents start having nonlinear agent-agent interactions, order and new structures appear. The new structure encodes certain information and memory, leading to reduced structural symmetry and higher complexity levels. *All complex systems possess memory. The more adaptive a complex system is, the more memory and order it captures in its structure.*

Additional Readings

Cowan, A. G., Pines D., and Meltzer D. (1994). "Complexity, Metaphors, Models, and Reality". Proceedings Volume XIX, Santa Fe Institute Studies in the Science of Complexity. Addison-Wesley Publishing Company. Reading, MA.

Meyers, R. A. (2009). "Encyclopedia of Complexity and Systems Science". Springer Science plus Business Media, LLC. New York, NY.

Pyka, A., and Scharnhorst A. (2009). "Innovation Networks, New Approaches in Modeling and Analyzing." Springer-Verlag. Heidelberg, Germany.

Gros, Claudius (2015). "Complex and Adaptive Dynamical Systems". Fourth Edition, Springer International Publishing AG, Switzerland.

Sayama, Hiroki (2015). "Introduction to the Modeling and Analysis of Complex Systems". Open SUNY Textbooks, Milne Library, Geneseo, NY.

Shayan, S. A. (2020). "Fundamental Constants in Mathematics & Physics – Are they universal codes?" Independent Publisher, Amazon.

Shayan, S. A. (2019). " Understanding Complex Adaptive Systems," Independent Publisher, Amazon.

Strogatz, S.H. (2015). "Nonlinear Dynamics and Chaos." Westview Press. Boulder, Colorado.

Manrubia, C. S., Mikhailov, A. S., and Zanette, D. H. (2004). "Emergence of Dynamical Order, Synchronization Phenomena in Complex Systems." World Scientific Lecture Notes in Complex Systems – Vol. 2, Singapore.

Thurner, S., Hanel, R., and Klimek, P. (2018). "Introduction to the Theory of Complex Systems." Oxford University Press. Oxford, UK.

About the Author

Shahin A. Shayan is a global investment & risk management consultant. By the end of 2016, he was the Hoda International Financial Engineering Company chairman, a private global investment banking/corporate finance advisory operation. Currently, he advises corporations on their socially responsible economic enterprises, complex financing setups, new venture structures, investments, corporate restructuring, enterprise-wide risk management frameworks, corporate valuation, and privatization issues related to the US and the Middle Eastern companies.

Initially, he worked as a young research scientist at NASA's Jet Propulsion Laboratories in Pasadena, CA, on Ion-Ion Scattering problems related to Voyager I & II projects. He later gained extensive financial experience in US firms such as Goldman Sachs & Co. in New York City, Security Pacific Merchant Bank, and First Interstate Bancorp in Los Angeles. He later gained valuable operational experience as an executive running investment operations related to volatile and uncertain developing environments in the Middle East, the challenge which he found very appealing. He innovatively structured and started two significant Orphan Funds by raising funds from the official Capital Markets in the Middle East, helping 20,000+ orphans in the regions.

He was born in Teaneck, New Jersey. Between 1976-84, he earned his B.A. in Quantum Chemistry, B.S. in Chemical Engineering, M.S. in Chemical Engineering, and an MBA from Columbia University in New York City. While working, he received his Doctorate Degree in Business Administration from California Coast University in 2014. He has completed five specialized Executive Management Programs at Harvard Business School in Boston, MA, throughout the years.

He has published six books and has been an active university lecturer and Executive Management Program speaker for 25 years. He has written many scholarly papers on a wide range of topics related to Complexity Theory, Complex Adaptive Systems, Fundamental Constants in Nature, Financial Engineering, Corporate Finance, Islamic Finance, Investment & Risk Management, Corporate Governance, Corporate Social Responsibility, and structuring Joint Social & Economic Enterprises, to name a few.

www.ingramcontent.com/pod-product-compliance
Lightning Source LLC
Chambersburg PA
CBHW052352220526
45465CB00003BA/1073